# Information Technology and Global Governance

Series Editor
Derrick Cogburn
American University
Bethesda, MD, USA

Information Technology and Global Governance focuses on the complex interrelationships between the social, political, and economic processes of global governance that occur at national, regional, and international levels. These processes are influenced by the rapid and ongoing developments in information and communication technologies. At the same time, they affect numerous areas, create new opportunities and mechanisms for participation in global governance processes, and influence how governance is studied. Books in this series examine these relationships and influences.

More information about this series at
http://www.palgrave.com/gp/series/14855

Nicola Palladino · Mauro Santaniello

# Legitimacy, Power, and Inequalities in the Multistakeholder Internet Governance

## Analyzing IANA Transition

palgrave
macmillan

Nicola Palladino
Department of Political and Social
Studies of the University of Salerno
Internet and Communication Policy
Centre
Fisciano, Salerno, Italy

Mauro Santaniello
Department of Political and Social
Studies of the University of Salerno
Internet and Communication Policy
Centre
Fisciano, Salerno, Italy

Information Technology and Global Governance
ISBN 978-3-030-56130-7        ISBN 978-3-030-56131-4   (eBook)
https://doi.org/10.1007/978-3-030-56131-4

Cover illustration: © Melisa Hasan

This Palgrave Macmillan imprint is published by the registered company Springer Nature
Switzerland AG
The registered company address is: Gewerbestrasse 11, 6330 Cham, Switzerland

# PRAISE FOR *LEGITIMACY, POWER, AND INEQUALITIES IN THE MULTISTAKEHOLDER INTERNET GOVERNANCE*

"Here is the enlightening survey of Internet governance and participatory democracy that everyone expects. The book assesses with lucidity the future of multistakeholderism as a solution to multilateralism in decline, which generates great hopes and sad disillusionment altogether. Combining a scholarly vision with a view from within the Internet community it mixes observation and theory to analyze the unending transition towards a self-regulating and decentralized decision-making system. While its promoters rhetorically see it as a benchmark for international activities beyond the Internet, the authors provide evidence that oligarchic power has not disappeared from the front stage. This is path breaking—an exhaustive and stimulating journey towards a more adequate and comprehensive knowledge of a vital sector that is too often victim of academic neglect."

—Yves Schemeil, *Professor of Political Science, University of Grenoble, France, and Professor of Global and Comparative Studies, at the Institut Universitaire de France*

"Palladino and Santaniello offer an important assessment of the legitimacy of the IANA transition process. They address both empirically and conceptually the essential shortcomings of multistakeholder governance. This book represents a valuable contribution to internet governance scholarship and to the study of multistakeholder governance in general."

—Jean-Marie Chenou, *Assistant Professor, Political Science, University of the Andes, Colombia*

# CONTENTS

1   Introduction: The IANA Transition and Internet
Multistakeholder Governance      1
     *1.1    Multistakeholderism in Internet Governance*      3
     *1.2    The Discursive Nature of Multistakeholderism:*
           *Performing Narrative or Misleading Rhetoric?*      9
     *1.3    Structure of the Book*      14
     *References*      16

2   Foundations, Pitfalls, and Assessment
of Multistakeholder Governance      21
     *2.1    Establishing Legitimate Authority for Transnational*
           *Governance: The Promise of Multistakeholderism*      22
     *2.2    Structural Pitfalls of Multistakeholder Governance*      26
     *2.3    A Framework to Assess the Legitimacy*
           *of Multistakeholder Initiatives*      31
     *References*      38

3   IANA Functions, ICANN, and the DNS War      43
     *3.1    The Domain Name System and IANA Functions*      43
     *3.2    From Technical Governance to Governance by Contract*      46
     *3.3    The Internationalization of the DNS*
           *and the Multistakeholder Model*      51

3.4    *The Evolution of ICANN Governance Structure*    53
*References*    58

**4    The Institutional Design of the IANA Transition Process**    63
4.1    *The Preparatory Phase*    64
4.2    *The Drafting of the IANA Stewardship Transition Proposal*    70
   4.2.1    *The Name Proposal*    71
   4.2.2    *The Number and Protocol Proposal*    74
   4.2.3    *The Assembled Proposal*    75
4.3    *The Drafting of the Enhancing ICANN Accountability Recommendations*    76
4.4    *Approval and Implementation by the ICANN Board and NTIA*    79

**5    The Input Legitimacy of the IANA Transition Process**    81
5.1    *Analyzing Input Legitimacy: Objectives, Data, and Methods*    82
5.2    *Inclusiveness and Balanced Representation*    86
5.3    *Representativeness, Revolving Doors, and Blurring Boundaries Among Stakeholders' Groups*    90
5.4    *Power Relationship in the "Small World" of the IANA Transition Network*    96
*References*    100

**6    The Throughput Legitimacy of the IANA Transition Process**    103
6.1    *The Procedural Quality of the IANA Transition*    104
6.2    *Discursive Quality*    110
   6.2.1    *Theoretical and Methodological Notes*    110
   6.2.2    *The Discussion on Post-transition Oversight Arrangements*    112
   6.2.3    *Evaluating Deliberativeness and Discourse Balance*    116
*References*    123

**7    The Output Legitimacy of the IANA Transition Process**    127
7.1    *Output Legitimacy in Constituent Policy-Making*    127

7.2   *Institutional Effectiveness: The New ICANN Bylaws*          129
7.3   *Outcome Effectiveness: A Still Contested DNS Regime*          135
    7.3.1   *National DNSs: The 2019 Russian Sovereign*
           *Internet Law*          135
    7.3.2   *The Territorial Jurisdiction of ICANN: The*
           *Case of the .org TLD*          137
*References*          141

8  **Conclusion: The Misleading Rhetoric
of Multistakeholderism**          143
*References*          156

# About the Authors

**Dr. Nicola Palladino** is a Research Fellow in the Department of Political and Social Studies of the University of Salerno, where he works at the Internet & Communication Policy Center. His main research interests concern Internet Governance, Internet Policy, and Digital Constitutionalism processes.

**Dr. Mauro Santaniello** is a Researcher in the Department of Political and Social Studies of the University of Salerno, where he teaches Internet Governance and Digital Policy. He is a co-founder and co-director of the Internet & Communication Policy Center. He has led several research groups working on Internet Governance.

# Abbreviations

| | |
|---|---|
| ALAC | Ad-Large Advisory Committee |
| ASO | Address Supporting Organization |
| ccNSO | Country Code Names Supporting Organization |
| DNS | Domain Name System |
| DoC | Department of Commerce |
| DoD | Department of Defense |
| GAC | Governmental Advisory Committee |
| GNSO | Generic Names Supporting Organization |
| IAB | Internet Architecture Board |
| IANA | Internet Assigned Numbers Authority |
| ICANN | Internet Corporation for Assigned Names and Numbers |
| IESG | Internet Engineering Steering Group |
| IETF | Internet Engineering Task Force |
| ISOC | Internet Society |
| ITU | International Telecommunication Union |
| NRO | Number Resource Organization |
| NTIA | National Telecommunications and Information Administration |
| PIR | Public Interest Registry |
| RIRs | Regional Internet Registries |
| RSSAC | Root Server System Advisory Committee |
| SSAC | Security and Stability Advisory Committee |
| WSIS | World Summit on the Information Society |

# LIST OF FIGURES

Fig. 5.1  Overlapping membership among IANA transition
         stakeholders (*Source* Authors' Creation)          95
Fig. 5.2  The IANA transition affiliation network (*Source* Authors'
         Creation)                                          97

# LIST OF TABLES

Table 2.1   Types, dimensions, and criteria of multistakeholder
            legitimacy                                              37
Table 5.1   Distribution of voting members among IANA/ICANN
            constituencies and sectorial categories                86
Table 5.2   Distribution of the 90 IANA transition members
            among geographical and geopolitical categories         89
Table 5.3   Composition of extended network                        93
Table 5.4   Overlapping rate between IANA constituencies
            and stakeholder types                                  94
Table 5.5   Affiliation network properties                         97
Table 5.6   Centrality measures for stakeholder categories         99

# CHAPTER 1

# Introduction: The IANA Transition and Internet Multistakeholder Governance

**Abstract** This chapter clarifies the purpose of the study, which is a critical assessment of the multistakeholder model in the Internet governance ecosystem through an in-depth analysis of the so-called Internet Assigned Numbers Authority (IANA) transition, probably the most relevant Internet governance multistakeholder process that has occurred in recent years. The chapter points out how multistakeholderism is a fuzzy concept that has led to ambiguous practices and disappointing results. Further, it highlights the discursive and legitimizing nature of multistakeholderism, which can serve both as a performing narrative capable of democratizing the Internet governance domain, as well as a misleading rhetoric solidifying the dominant position of the most powerful actors in different Internet policy-making arenas. Finally, the chapter concludes that a deep investigation of the consistency of the IANA transition process with normative standards of democratic legitimacy for transnational governance could shed light on the evolution of multistakeholderism in this field.

**Keywords** IANA transition · Internet governance · Multistakeholderism · Legitimacy

N. Palladino and M. Santaniello, *Legitimacy, Power, and Inequalities in the Multistakeholder Internet Governance*, Information Technology and Global Governance, https://doi.org/10.1007/978-3-030-56131-4_1

On March 14, 2014, the US government officially announced its intention "to transition key Internet domain name functions to the global multistakeholder community," (NTIA 2014) relinquishing the special role it previously carried out in the management of the domain name system (DNS).

The decision-making process following this announcement, commonly referred to as the IANA transition, was a milestone in the history of Internet governance, yet curiously it is still under-investigated.

IANA functions are the key tasks of the DNS, providing all devices and systems connected to the Internet with consistency between IP addresses and domain names. Historically managed by the technical community of engineers and academics that created the Internet, in 1998 these functions were transferred to a newly designed organization, ICANN, established as a not-for-profit corporation under California law. The authority of ICANN over the DNS has been contested since the beginning, both because of the dominant role of Western businesses and its relationship with the US government, which reserved for itself a special oversight role (Froomkin 2000; Bygrave 2015; Calandro et al. 2013; Weber and Gunnarson 2012).

The NTIA announcement raised the expectation that these long-standing controversies would be fixed. In the announcement, the multistakeholder model of governance was the key concept guiding the IANA transition process. Not only the IANA transition was supposed to be a multistakeholder effort, in NTIA's intention it was even expected to "support and enhance the multistakeholder model of Internet policymaking and governance." Indeed, for many actors it represented the highest implementation of the concept of multistakeholderism within the Internet governance field.

This book explores the extent to which the IANA transition succeeded in establishing a real multistakeholder model for the governance of the DNS and the meaning of the IANA transition in the development of Internet multistakeholder governance.

To fulfill this purpose, it is necessary to briefly retrace the history of multistakeholderism within Internet governance, its meaning, and how the IANA transition is placed within its evolution.

## 1.1 MULTISTAKEHOLDERISM IN INTERNET GOVERNANCE

The concept of multistakeholder governance appeared in the Internet governance debate during the process that brought to the World Summit on Information Society (WSIS), a UN summit convened by the International Telecommunications Union (ITU), which took place in two stages: a first summit was held in Geneva from 10 to 12 December 2003, while the second part of the initiative was carried out in Tunis from 16 to 18 November 2005.

First references to a "multistakeholder approach" emerged during the preparatory phase as a middle ground between different positions. On the one side, Western countries and business communities supported the status quo and conceived Internet governance as limited to the technical management of DNS, to be carried out through a regime of private self-regulation with the oversight of the US government. On the other side, several actors contested this view, fostering a broader conception of Internet governance that includes public policy issues. These actors included developing countries that sought the sovereignty of states over the Internet (including DNS) to be exercised through traditional intergovernmental arrangements, as well as transnational civil society calling for a more transparent, accountable, and human rights-based development of the Internet (Mueller 2010; Hofmann 2007; Radu 2019).

The term "multistakeholder" officially entered the language of Internet governance with the establishment of the Working Group on Internet Governance (WGIG) (Kummer 2013). The WGIG was an expert group set up between the two conferences to circumvent the abovementioned political deadlock. It was asked to provide a shared definition of Internet governance, identify the public policy issues that are relevant to this field, and propose a common understanding of the roles and responsibilities of the different stakeholder groups. "Multistakeholder governance" was one of the key concepts in the WGIG final report, which was incorporated into the final output of the summit, the Tunis Agenda.

The WSIS led to the definition of Internet governance as "the development and application by governments, the private sector and civil society, in their respective roles, of shared principles, norms, rules, decision-making procedures, and programmes that shape the evolution and use of the Internet" (WGIG 2005). Moreover, the WSIS recognized that Internet governance also includes public policy issues under the responsibility of states, which nevertheless are excluded from the day-to-day technical and operational matters entrusted to the private sector. Yet, the WSIS outcomes acknowledged an important role for civil society, academic and international organizations, even though their roles remained roughly approximated.

This definition of Internet governance constitutes a milestone in the history of Internet governance. Despite its vagueness, it represents the full acknowledgment by a broad and variegated Internet community, including governments and intergovernmental organizations, that the complexity of Internet governance does not allow the leadership of a single class of actors, but rather requires decentralized, bottom-up, inclusive policy development, and decision-making processes with the collaboration of all affected parties (Doria 2014; Kleinwächter 2011; Zingales and Radu 2015; Chenou and Radu 2014).

Nevertheless, the WSIS produced no substantial changes concerning the concrete mechanisms of governance. ICANN maintained its role in the DNS management and the questions of its jurisdiction and its special relationship with the USA were not mentioned in the final document.

The greater institutional innovation was the establishment of the Internet Governance Forum (IGF), designed as a "forum for multi-stakeholder policy dialogue" to be held every year for a period of five years (then renewed in 2005 and 2010). As paragraph 77 of the Tunis Agenda clarifies, the IGF "would not replace existing arrangements, mechanisms, institutions or organizations, but would involve them and take advantage of their expertise. It would be constituted as a neutral, non-duplicative and nonbinding process. It would have no involvement in day-to-day or technical operations of the Internet" (WSIS 2005). As noted, the creation of the IGF was "the kind of agreement that could get the WSIS out of its impasse; it allowed the critics to continue raising their issues in an official forum, but as a nonbinding discussion arena, could not do much harm to those interested in preserving the status quo" (Mueller 2010: 78; see also Becker 2019).

By and large, the outcomes of the WSIS process resulted in a compromise between incompatible views (Mansell 2007; Musiani and Pohle 2014; Raymond and DeNardis 2015), which did not produce a synthesis. As noted, "rather than cutting short the issues, the WSIS promoted further dialogue" (Chenou and Radu 2014: 8).

It should be noted that the ambiguous results of the WSIS led to the equally ambiguous development of multistakeholderism within the Internet governance ecosystem. Following the WSIS, the concept and practices of multistakeholder governance have rapidly spread across many Internet governance contexts, and a large number of Internet-related organizations, arenas, and processes started to describe themselves as multistakeholder (Chenou and Radu 2014; Doria 2014; Raymond and DeNardis 2015). This is largely because multistakeholder processes provide some guiding principles and common understandings in a chaotic and fragmented policy domain fraught with tensions. As noted, "the multi-stakeholder concept is a discursive artefact that aims to smooth contradictory and messy practices into a coherent story about collaborative transnational policymaking" (Hofmann 2016: 44). As we will see in more detail later, "multistakeholder" has become a legitimizing label under which most of the actors within the Internet governance ecosystem have tried to place themselves in order to justify and legitimate their claims.

This development has been helped by the "creative ambiguity" of the Tunis Agenda (Kummer 2013), as well as by the "fuzzy" nature (Cammaerts 2011) and the under-theorization of the multistakeholder concept (Cammaerts 2011; Zingales and Radu 2015; Raymond and DeNardis 2015). This has fostered the proliferation of different interpretations and forms of multistakeholder arrangements "susceptible to use in attempts to conceal or advance particular interests or agendas" (Raymond and DeNardis 2015: 2).

On the other hand, the outcomes of the WSIS left the tensions surrounding Internet governance unresolved, giving rise to contestation in subsequent years and to the cyclical recurrence of political conflicts challenging the consensus around the multistakeholder model.

The IGF was criticized for its strict focus on dialogue and lack of any decision-making power, even in the "soft" form of official statements, guidelines, or recommendations. Thus, the global forum has been defined as a "talk-shop initiative" (Zittrain 2008; Mueller 2010) giving rise to nothing more than "bland consensus pronouncements instead of getting at the 'nuts and bolts' of the problems it identifies" (Van Eeten and Mueller 2012: 727). This discontent regarding the poor impact of the IGF in 2010 led China and the Group of 77 to threaten to not renew its mandate during the review process (Brousseau et al. 2012; Santaniello 2016).

The multistakeholder model seemed to receive a set-back during the World Conference on International Communications (WCIT) in 2012.

The conference was convened by the ITU to revise and update the 1988 International Telecommunication Regulations (ITRs), a global treaty providing international telecommunication services and networks with general principles for interconnection and interoperability, in light of developments in the telecommunications system. As Weber pointed out "the potential implementation of multistakeholderism by the WCIT has been partly perverted" (Weber 2014: 101) both in its practices and contents. The conference was criticized for discussing the most important motions in closed rooms; for reserving voting power only to states, preventing the meaningful participation of other stakeholders; and for deviating from the traditional ITU consensus rule, allowing a majoritarian vote on the outcome. Indeed, for the first time, Western countries were defeated by a coalition of "sovereigntist" developing countries led by China and Russia. Although the final text reaffirmed the commitment toward multistakeholder governance as settled in the Tunis Agenda and did not include more radical proposals submitted during the discussion, the outcomes of the WCIT have commonly been interpreted as an attempt to shift Internet governance toward an intergovernmental model, or at least toward a re-defined multistakeholder model where states play a leading role (Chenou and Radu 2014; Weber 2014; Radu 2019). From a different point of view, the WCIT made clear that the Internet governance status quo was no longer supported by international consensus and that unresolved issues could be no longer circumvented without threatening the very existence of a *global* Internet, since several states started to prefer a fragmented scenario to obtain greater control over their information systems (Santaniello 2016; Rioux et al. 2014; Hill 2014).

The next year, Edward Snowden's disclosures about a mass surveillance program carried out by the US National Security Agency (NSA), with the collusion of big Internet companies such as Verisign, Skype, and Google (Greenwald 2014), "further weakened the US position in shaping the future of the Internet" (Becker 2019: 9) and undermined the "trust of the global community in its role as caretaker of the Internet" (Cogburn 2017: 20). It should be said that the greatest impact of this scandal was in breaking up the traditional dominant coalition revolving around the leading role of the US government and its industries (Mueller and Wagner 2014). Indeed, actors like the EU and many civil society organizations had previously supported the coalition mainly for the sake of Internet integrity, openness, and freedom, considering it the lesser of two evils compared with an intergovernmental model favoring the positions of authoritarian countries. For this reason, although mass surveillance activities did not involve DNS operations (DeNardis 2014), the Snowden scandal undermined the authority of ICANN and led to calls for the relinquishment of the USA's special role in its governance.

However, several actors, including a large part of the ICANN community itself, saw in this crisis an opportunity to boost multistakeholder governance on a different basis. In October 2013, Brazilian President Dilma Rousseff announced a high-level multistakeholder meeting on the future of Internet governance, later called NETmundial, to be held in Brazil in April 2014, to "reimagine what the Internet governance ecosystem could look like and to formally submit recommendations for new principles and values for this global community" (Cogburn 2017: 20). In the same month, the leaders of ICANN and the other organizations responsible for the coordination of the Internet's technical infrastructure released the Montevideo Statement, in which, after warning against fragmentation and loss of confidence in the Internet field, they called for catalyzing "community-wide efforts towards the evolution of global multistakeholder Internet cooperation" and "accelerating the globalization of ICANN" toward "an environment in which all stakeholders, including all governments, participate on an equal footing" (ICANN et al. 2013). Similarly, in February 2014, the European Commission, after condemning "large-scale Internet surveillance" for reducing trust in the Internet and leading to fragmentation, called for "the globalization of the ICANN and the IANA functions" and renewed its support for "a *real* multi-stakeholder governance model for the Internet based on the full involvement of all relevant actors and organizations" (EU-COM 2014, emphasis added).

Such external pressure finally led a reluctant US government to relinquish its special role in the DNS (Becker 2019; Post and Kehl 2015) to "prevent states with opposing preferences like China and Russia from being granted greater control over the DNS" (Becker 2019: 10) within some intergovernmental arrangement under the UN system.

The IANA transition thus represented a challenge for the multistakeholder model in the Internet governance ecosystem. On the one hand, its success was believed to "bolster the multistakeholder approach to Internet governance and offer valuable lessons to inform a broader range of governance decisions" (Post and Kehl 2015: 28). On the other hand, in the words of Jonathan Robinson, who chaired one of the groups charged with the drafting of the transition proposal, if the "process failed, it might expose a fatal weakness in the multi-stakeholder movement" (Robinson 2016: 201).

In the end, the IANA transition process concluded its work in 2016 and succeeded in removing the US government's oversight. In Robinson's view, this means that "the whole multi-stakeholder process was put to the test and seems to have passed, which is good news" and after the transition the multistakeholder model "does provide something unique, and something that has actively evolved into something more mature" (Robinson 2016: 201–203). According to the numerous statements in support of the IANA transition proposal, this is a shared opinion within the Internet community that goes beyond the predictable support coming from more directly involved subjects like ICANN, NTIA, and IETF. Several actors in the private sector and civil society consider the IANA transition a turning point in the empowerment of the multistakeholder model of governance and the Internet as a whole.

Nevertheless, it is worth noting that the most recurring argument in support of this thesis is that the removal of US oversight weakens the intergovernmental position. The "Civil Society Statement of Support for the IANA Transition" makes the point clearly, affirming that "the transition of these functions away from the US government removes an excuse for authoritarian countries to demand greater oversight and regulation of Internet issues" and thus "this transition is returning the Internet and DNS to the open multistakeholder governance model that characterized and fostered its first few decades of growth."[1]

The weak point of this argument is that it relies on the conflation of the multistakeholder approach with the privatization of Internet governance. Not by chance, the NTIA announcement recalls the Commerce Department's Statement of Policy (June 10, 1998) stating that the US government "is committed to a transition that will allow the private sector to take leadership for DNS management." The concept was reaffirmed in the NTIA's final assessment of the IANA Stewardship Transition Proposal, which was deemed to meet "the criteria necessary to complete the long-promised privatization of the IANA functions" (NTIA 2016). Even if in this context the term "private" is employed with a broad meaning indicating non-state actors, multistakeholder governance is supposed to be something more than putting together different types

---

[1] https://www.accessnow.org/cms/assets/uploads/2016/05/CSstatementonIANAtransitionMay2016-1.pdf. Accessed 12 June 2020.

of actors to avoid the leading role of governments in addressing an issue. As we will see in more detail later, multistakeholderism calls into question principles such as deliberative democracy, equity, participation, representativeness, and accountability.

These last considerations make clear that the evaluation of the IANA transition's contribution to the development of multistakeholderism within the field of Internet governance requires deepening the understanding of what multistakeholderism means and what it ought to be from a normative point of view.

Despite the pivotal importance commonly assigned to the IANA transition, studies assessing how this process conformed with normative standards of multistakeholder governance are still missing. This book fills this gap by providing a theoretical framework and empirical research to analyze the extent to which the IANA transition has been a real multistakeholder process that gave rise to novel and higher multistakeholder arrangements for the governance of the DNS.

## 1.2   The Discursive Nature of Multistakeholderism: Performing Narrative or Misleading Rhetoric?

The previous discussion highlighted how multistakeholder governance has followed a contradictory development within the Internet governance ecosystem. It has spread widely, becoming a benchmark within the Internet governance ecosystem, while at the same time its concrete implementations have given rise to disappointment and contestation. It is necessary then to explore this ambiguous dynamic to better understand the meaning of the multistakeholder model in the governance of the Internet, as well as to better place the IANA transition within the history of Internet (multistakeholder) governance.

Janette Hofmann recently addressed the issue of the discrepancy between the "rising popularity" and "well-known performance problems" (Hofmann 2017: 1) of multistakeholder governance in the Internet field, by adopting a discursive approach. She builds upon the definition of discourse elaborated by Hajer, according to which discourses are "specific ensemble of ideas, concepts, and categorizations that are produced, reproduced, and transformed in a particular set of practices and through which meaning is given to physical and social realities" (Hajer 1995: 44), organized around storylines, that is "narratives on social reality [...] that provide actors with a set of symbolic references that suggest a common understanding" (ibidem, 62).

In her view, multistakeholderism is a narrative based on three main promises: the promise of achieving global representation on an issue putting together all the affected parties; the promise of overcoming the traditional democratic deficit at the transnational level, "establishing communities of interest as a digitally enabled equivalent to territorial constituencies" (Hofmann 2017: 4); and the promise of higher and enforced outcomes since incorporating global views on the matter through a consensual approach should ensure more complete solutions and their smooth implementation. In the end, these promises could be summarized in a "meta-narrative" according to which "implementing principles such as inclusiveness, transparency, equality and procedural fairness, the national concept of democracy can be extended beyond territorial borders and thereby confer to transnational policy making the legitimacy it still lacks" (Hofmann 2017: 5).

"Legitimacy" then is the key concept for understanding the spread and relevance of multistakeholderism in the Internet governance field. As we will see in more detail in Chapter 2, the authority of an institution or an arrangement at the transnational level cannot be ensured through legal/coercive enforcement capacity; rather, rule compliance might rely on legitimacy, understood as the consent of the rule-addressees, leading to voluntary cooperation (Hurd 1999; Beisheim and Dingwerth 2008; Tallberg et al. 2018)

This is true also with regard to global Internet governance and its institutions (Take 2012), and in particular for ICANN, which since the beginning has suffered from a lack of meaningful legitimacy (Malcolm 2008; Weinberg 2000) due to the conditions of its establishment (see Chapter 3). Indeed, ICANN never received full international recognition, particularly in the UN system (Mueller et al. 2007), nor was it established with the support of all affected non-state actors, including some parts of the technical community (Mueller 1999, 2002).

Multistakeholderism has thus been conceived as a viable solution to the problem of legitimacy beyond the nation-state, capable of providing an alternative form of democratic legitimacy suitable for transnational policy-making (Bäckstrand 2006; Macdonald 2008) and for this reason it was embraced with great enthusiasm and expectation within the Internet governance field (Cammaerts 2011; Zingales and Radu 2015; Hintz 2007; Weinberg 2000, 2011; Mueller and Wagner 2014).

The symbolic value of multistakeholderism is so powerful that, according to Hofmann (2016, 2017), it goes beyond its effective participatory and regulatory capacity. In her view, multistakeholderism is a "performative" narrative that does not represent reality but is nevertheless capable of shaping the development of Internet governance toward specific political goals and aspirations. This conception assumes, on the one hand, the existence of an insurmountable gap between the multistakeholder model of governance as an ideal-type and its concrete implementations, and, on the other hand, that actors make efforts to comply with the normative standards prescribed by such an ideal-type.

Nevertheless, the conception of multistakeholderism as a legitimizing discourse or narrative opens the door to other interpretations of the development of multistakeholder governance in the Internet domain. As Hajer found in his studies on environmentalism and sustainable development, when a narrative is vague enough to collect around itself a wide range of actors with different cognitive and political commitments it can serve as a "symbolic umbrella" under which different discourse-coalitions engage in a "conflict of interpretation," struggling around the basic terms of the narrative to achieve hegemony in that policy domain (Hajer 1995: 14–15).

Something similar seems to have happened in the Internet field with multistakeholderism. Several studies have detected different discursive or ideological coalitions in the Internet governance field (Mueller 2010; Chenou 2014; Santaniello and Palladino 2017). Despite the differences, each of these works identifies a neoliberal discourse, supported by many Western governments and business interests, according to which the multistakeholder model is "a way to improve self-regulation" policies; a sovereigntist discourse shared by BRICS and developing countries which conceive of the multistakeholder approach as a more open variant of the intergovernmental model; and a discourse promoted by civil society that has "interpreted multi-stakeholderism as participatory democracy" (Radu 2019: 101).

In the end, as many other scholars have observed, the several forms of multistakeholderism in the Internet governance field could be reduced to two main conceptions (Doria 2014; Koechlin and Calland 2009; Zingales and Radu 2015): the first grounded in a belief in the full and equal participation of all concerned stakeholders in decision-making processes; the second foreseeing the leading role of one or few stakeholder groups, with governance processes open to discussing with other stakeholders who are devoid of real decisional power.[2] According to Avri Doria (2014), this second interpretation is an "abuse" of the term multistakeholderism "used by single stakeholder groups who exclude other stakeholders in an attempt to mask the organization's single stakeholder nature" (Doria 2014: 115). From a slightly different perspective, this second interpretation implies a conception of multistakeholderism as a rhetorical device employed to legitimize the domination of a particular group of actors over the others.

Looking at the existing literature, the vast majority of the multistakeholder practices in the Internet governance field should be considered an abuse of the term. Criticisms about the unequal participation of stakeholders, in which some groups enjoy structural supremacy and influence over decision-making processes, have been raised toward the multistakeholder initiatives promoted by intergovernmental organizations such as the ITU, WSIS, IGF, and WCIT (Hintz 2007; Cammaerts 2011; Padovani 2012; Doria 2014; Weber 2014; Raymond and DeNardis 2015) as well as toward ICANN (Malcolm 2008; Antonova 2008; Chenou 2014; Carr 2015; Calandro et al. 2013; Froomkin 2000).

As a whole, a common concern of observers is the risk that multistakeholderism may result in a "rhetorical exercise aimed at neutralising criticism" (Padovani and Pavan 2007: 100) without actually giving up power (see also Cammaerts 2011; Padovani 2012). Carr (2015: 658) added that "multistakeholderism reinforces existing power dynamics that have been 'baked in' to the model from the beginning. It privileges north-western governments, particularly the US, as well as the US private sector." Similarly, Chenou (2014) defined multistakeholderism as a discursive tool employed to create consensus around the hegemony of a power élite.

Nevertheless, researchers also underline the ongoing efforts that some organizations, such as ICANN and the IGF, have been undertaking to reply to these criticisms (Hofmann 2016; Radu 2019), in line with the conception of multistakeholderism as a "performative" narrative.

---

[2] Similarly, Fransen and Kolk (2007) in their account of multistakeholder practices outside Internet governance distinguished between *involvement*, based on the broad inclusiveness of all stakeholder in the decision-making process, and *consultation*, where some stakeholder groups play only an advocacy role without voting rights.

We believe that all the described perspectives represent some aspects of the development of multistakeholderism in the Internet governance ecosystem. Multistakeholderism is a legitimizing discourse that spur actors to comply with democratic standards for global governance, or at least to claim to do so. Equally, many actors and institutions adhere to multi-stakeholderism to acquire legitimization for their dominant position or agenda. They exploit the ambiguity and structural deficits of the multi-stakeholder framework in their favor, re-defining to their advantage the basic terms of multistakeholderism in order to exert hegemony in some specific policy-arena.

Multistakeholder initiatives in the Internet governance field are thus constantly exposed to the risk that their legitimizing power could be perverted toward the legitimation of a power élite that exerts a major influence on the process and is better placed to safeguard its particular interest.

In this regard, it is useful to recall the concepts of "discourse institutionalism" (Schmidt 2008, 2010) and "discursive institutionalization" (Hajer 1993, 1995), according to which discourses create and maintain institutions insofar as they solidify into routinized practices and approaches. However, as Hajer warns, this process does not occur in a "social vacuum," rather it is affected by already existing institutional arrangements and the discourses and power positions embedded in them.

This is particularly true with regard to a vague discourse such as multistakeholderism, which acts as a "symbolic umbrella" under which definitional struggles take place. It is not surprising, then, that when a multistakeholder approach is adopted within an already existing institution like ICANN or the ITU, it is reshaped according to previously consolidated practices, categorizations, and routines, favoring the dominant actors in each arena.

These considerations suggest that answering if multistakeholderism in Internet governance represents a genuine attempt at democratizing this transnational issue domain, or if it results in misleading rhetoric legitimizing already existing power inequalities requires a deep investigation of the major multistakeholder initiatives in the field and taking into account the consistency of their concrete governance arrangements with specific normative standards, as well as the institutional background of each initiative.

The IANA transition is the latest milestone in the history of the multistakeholder governance of the Internet. Therefore, it constitutes an emblematic case and a good starting point for investigating if meaningful advancements have been made toward more participatory governance of the DNS, or if the multistakeholder model merely served hegemonic practices reproducing and solidifying already existing power imbalances. Assessing the degree of "multistakeholderness" means exploring the extent to which both the decision process and its outcome succeeded in fixing the well-known governance limitations at the basis of ICANN legitimacy deficit, leading toward a system where all the components of the multistakeholder global Internet community could meaningfully influence the management of the DNS.

## 1.3    STRUCTURE OF THE BOOK

Chapter 2 deepens the reflection on multistakeholderism, pointing out how events in the Internet governance field followed a more general trend in global governance studies and practices aiming at providing transnational policy-making with novel democratic legitimacy standards. We stress how multistakeholderism attempts to establish legitimate authority at the global level, replacing the legitimacy derived from electoral mechanisms with one coming from an inclusive deliberative process. Further, we highlight the structural weaknesses of the multistakeholder model. In particular, we maintain that the deliberative and consensual approach at the basis of the multistakeholder model overlooks the dimension of power, especially in its institutional and discursive forms, giving the most powerful actors the chance to steer the process toward their preferred outcomes.

Finally, through a literature review on the categories of input, throughput, and output legitimacy, the chapter identifies a set of legitimacy standards that a multistakeholder initiative should satisfy in order to fulfill the promises of multistakeholderism and avoid being considered just a rhetorical exercise masking practices of domination.

Chapter 3 provides the reader with the necessary case-background for understanding the content and issues at stake within the IANA transition process. First, we describe the distributed but hierarchical architecture of the DNS, the IANA functions, and why they are core elements of the DNS. Then we move our attention to the governance of the DNS, retracing the shift from the so-called technical regime toward "self-regulation" with the establishment of ICANN. In so doing, we focus on ICANN's governance structure and the controversies that undermined its legitimacy and its authority over the DNS. While the previous chapters set out the theoretical framework guiding our analysis, the following chapters address the empirical side of our inquiry. In Chapter 4, we describe the institutional design of the IANA transition process. The architecture of the process is traced back, highlighting how the drafting work was distributed among different working groups and outlining the decision-making chain that led to the final approval.

Chapter 5 explores the consistency of the IANA transition process, with the principle of input legitimacy broken down into the dimensions of "inclusiveness," the extent to which all the categories of actors affected by the issue were involved in the decision-making process; "balanced representation," taking into account the distribution of seats among stakeholder groups in decision-making bodies; and "representativeness," which consider if the participants at the IANA transition were genuine representatives of the respective constituencies.

Chapter 6 focuses on throughput legitimacy, namely it considers the process through which the participation of different stakeholders gave rise to the outcome. We analyze the procedural quality of the process in order to assess if the institutional design gave all involved actors equal and meaningful opportunities to participate and influence the final outcome. Further, we take into account the discursive quality of the process, considering the extent to which the deliberation was open, rational, and inspired by mutual respect and constructive spirit, as well as if the process was flawed by hegemonic discursive practices inhibiting minority points of view.

Chapter 7 assesses the IANA transition process from the perspective of output legitimacy, investigating whether the outputs of the process succeeded in transferring the stewardship of IANA functions to the multistakeholder global Internet community. Further, the chapter assesses whether the new arrangements and the bylaw's modifications made it possible to remedy the well-known limits of ICANN governance.

Finally, in Chapter 8 we draw some conclusions from the conducted analysis. Besides an overall assessment of the normative legitimacy of the IANA transition process, we consider what the experience could teach us about the development of multistakeholderism within the Internet governance field and its future prospects.

# References

Antonova, S. (2008). Deconstructing an Experiment in Global Internet Governance: The ICANN Case. *International Journal of Communications Law & Policy, 12*(1), 2–15.

Bäckstrand, K. (2006). Multi-stakeholder Partnerships for Sustainable Development: Rethinking Legitimacy. *Accountability and Effectiveness. European Environment, 16*(5), 290–306.

Becker, M. (2019). *When Public Principals Give Up Control over Private Agents: The New Independence of ICANN in Internet Governance.* Regulation and Governance.

Beisheim, M., & Dingwerth, K. (2008, June). *Procedural Legitimacy and Private Transnational Governance: Are the Good Ones Doing Better?* (Report). SFB Research Center, Freie Universität Berlin, DE. https://www.sfb-govern ance.de/publikationen/sfb-700-working_papers/wp14/SFB-Governance-Working-Paper-14.pdf. Accessed 18 March 2020.

Brousseau, E., Marzouki, M., & Méadel, M. (Eds.). (2012). *Governance, Regulation and Powers on the Internet.* Cambridge: Cambridge University Press.

Bygrave, L. A. (2015). *Internet Governance by Contract.* Oxford: Oxford University Press.

Calandro, E., Gillwald, A., & Zingales, N. (2013). *Mapping Multistakeholderism in Internet Governance: Implications for Africa.* https://ssrn.com/abstract=2338999. Accessed 14 September 2019.

Cammaerts, B. (2011). Power Dynamics in Multi-stakeholder Policy Processes and Intra-civil Society Networking. In R. Mansell & M. Raboy (Eds.), *The Handbook of Global Media and Communication Policy* (pp. 131–146). Oxford: Wiley-Blackwell.

Carr, M. (2015). Power Plays in Global Internet Governance. *Millennium: Journal of International Studies, 43*(2), 640–659.

Chenou, J. M. (2014). From Cyber-Libertarianism to Neoliberalism: Internet Exceptionalism, Multi-stakeholderism, and the Institutionalisation of Internet Governance in the 1990s. *Globalizations, 11*(2), 205–223. https://doi.org/10.1080/14747731.2014.887387.

Chenou, J. M., & Radu, R. (2014). Global Internet Policy: A Fifteen-Year Long Debate. In R. Radu, J. M. Chenou, & R. Weber (Eds.), *The Evolution of Global Internet Governance* (pp. 3–22). Berlin: Springer.

Cogburn, D. L. (2017). *Transnational Advocacy Networks in the Information Society.* New York: Palgrave.

DeNardis, L. (2014). *The Global War for Internet Governance.* New Haven: Yale University Press.

Doria, A. (2014). Use [and Abuse] of Multistakeholderism in the Internet. In R. Radu, J. M. Chenou, & R. Weber (Eds.), *The Evolution of Global Internet Governance* (pp. 115–140). Berlin: Springer.

EU-COM. (2014). *Commission to Pursue Role as Honest Broker in Future Global Negotiations on Internet Governance.* European Commission Press release. https://ec.europa.eu/commission/presscorner/detail/en/IP_14_142. Accessed 19 March 2020.

Fransen, L. W., & Kolk, A. (2007). Global Rule-Setting for Business: A Critical Analysis of Multi-stakeholder Standards. *Organization, 14*(5), 667–684.

Froomkin, A. M. (2000). Wrong Turn in Cyberspace. *Duke Law Journal, 50*(17), 17–184.

Greenwald, G. (2014). *No Place to Hide.* New York: Penguin.

Hajer, M. A. (1993). Discourse Coalitions and the Institutionalisation of Practise. In F. Fischer & J. Forester (Eds.), *The Argumentative Turn in Policy Analysis and Planning* (pp. 43–76). Durham and London: Duke University Press.

Hajer, M. A. (1995). *The Politics of Environmental Discourse.* New York: Oxford University Press.

Hill, R. (2014). The Internet, Its Governance, and the Multi-stakeholder Model. *Info, 16*(2), 16–46. https://doi.org/10.1108/info-05-2013-0031.

Hintz, A. (2007). *Deconstructing Multi-stakeholder: The Discourses and Realities of Global Governance at the World Summit on the Information Society (WSIS).* http://citeseerx.ist.psu.edu/viewdoc/download;jsessionid=26199B95118C AF71D7EED700354EDC33?doi=10.1.1.408.5532&rep=rep1&type=pdf. Accessed 19 March 2020.

Hofmann, J. (2007). Internet Governance: A Regulative Idea in Flux. In R. K. J. Bandamutha (Ed.), *Internet Governance: An Introduction* (pp. 74–108). Icfai: University Press.

Hofmann, J. (2016). Multi-stakeholderism in Internet Governance: Putting a Fiction into Practice. *Journal of Cyber Policy, 1*(1), 29–49.

Hofmann, J. (2017). *The Multi-stakeholder Concept as Narrative: A Discourse Analytical Approach.* https://papers.ssrn.com/sol3/papers.cfm?abstract_id=3070583. Accessed 20 September 2017.

Hurd, I. (1999). Legitimacy and Authority in International Politics. *International Organization, 53*, 379–408.

ICANN et al. (2013). *Montevideo Statement on the Future of Internet Cooperation.* https://www.icann.org/news/announcement-2013-10-07-en. Accessed 12 March 2020.

Kleinwächter, W. (2011). *Editorial* (MIND Paper Series, No. 2). Berlin: Internet & Gesellschaft Collaboratory.

Koechlin, L., & Calland, R. (2009). Standard Setting at the Cutting Edge: An Evidence-Based Typology for Multi-stakeholder Initiatives. In A. Peters, et al. (Eds.), *Non-state Actors as Standard Setters.* Cambridge: Cambridge University Press.

Kummer, M. (2013). *Multistakeholder Cooperation Reflections on the Emergence of a New Phraseology in International Cooperation*. https://www.internets ociety.org/blog/2013/05/multistakeholder-cooperation-reflections-on-the-emergence-of-a-new-phraseology-in-international-cooperation/. Accessed 12 March 2020.

Malcolm, J. (2008). *Multi-stakeholder Governance and the Internet Governance Forum*. Sidney: Terminus Press.

Mansell, R. (2007). Great Media and Communication Debates: WSIS and the MacBride Report. *Information Technologies & International Development, 3*(4), 15–36.

Macdonald, T. (2008). *Global Stakeholder Democracy, Power and Representation Beyond Liberal States*. Oxford: Oxford University Press.

Mueller, L. M. (1999). ICANN and the Internet Governance. *Info, 1*(6), 497–520.

Mueller, L. M. (2002). *Ruling the Root: Internet Governance and the Taming of Cyberspace*. Cambridge and London: MIT Press.

Mueller, L. M. (2010). *Networks and States: The Global Politics of Internet Governance*. Cambridge: MIT Press.

Mueller, L. M., Mathiason, J., & Klein, H. (2007). The Internet and Global Governance: Principles and Norms for a New Regime. *Global Governance, 13*, 237–254.

Mueller, L. M., & Wagner, B. (2014). *Finding a Formula for Brazil: Representation and Legitimacy in Internet Governance*. Internet Policy Observatory. https://global.asc.upenn.edu/app/uploads/2014/09/Finding-a-Formula-for-Brazil-Representation-and-Legitimacy-in-Internet-Governance.pdf. Accessed 12 March 2020.

Musiani, F., and Pohle, J. (2014). NET Mundial: Only a Landmark Event If 'Digital Cold War' Rhetoric Abandoned. *Internet Policy Review, 3*(1). https://doi.org/10.14763/2014.1.251.

NTIA. (2014). *Intent to Transition Key Internet Domain Name Functions*. https://www.ntia.doc.gov/press-release/2014/ntia-announces-intent-transition-key-internet-domain-name-functions. Accessed 12 June 2020.

NTIA. (2016). *IANA Stewardship Transition Proposal Assessment Report*. https://www.ntia.doc.gov/files/ntia/publications/iana_stewardship_transit ion_assessment_report.pdf. Accessed 10 March 2020.

Padovani, C. (2012). Democracy and Global Governance: The Wager of the Internet Governance Forum. In F. Massit-Folléa, C. Méadel, & L. Monnoyer-Smith (Eds.), *Normative Experience in Internet Politics*. Paris: Presses des Mines.

Padovani, C., & Pavan, E. (2007). Diversity Reconsidered in a Global Multi-stakeholder Environment: Insights from the Online World. In W. Kleinwachter (Ed.), *The Power of Ideas: Internet Governance in a Global Multi-stakeholder Environment* (pp. 99–109). Berlin: Marketing fur Deutschland GmbH.

Post, D. G., & Kehl, D. (2015). *Controlling Internet Infrastructure: The 'IANA Transition' and Why It Matters for the Future of the Internet, Part 1.* https://static.newamerica.org/attachments/2964-controlling-internet-infras tructure/IANA_Paper_No_1_Final.32d31198a3da4e0d859f989306f6d480. pdf. Accessed 15 September 2018.

Radu, R. (2019). *Negotiating Internet Governance.* Oxford: Oxford University Press.

Raymond, M., & DeNardis, L. (2015). Multistakeholderism: Anatomy of an Inchoate Global Institution. *International Theory.* https://doi.org/10.1017/s1752971915000081.

Rioux, M., Adam, N., & Perez, B. C. (2014). Competing Institutional Trajectories for Global Regulation—Internet in a Fragmented World. In R. Radu, J. M. Chenou, & R. Weber (Eds.), *The Evolution of Global Internet Governance* (pp. 37–56). Berlin: Springer.

Robinson, J. (2016). IANA Transition. *Journal of Cyber Policy, 1*(2), 198–205. https://doi.org/10.1080/23738871.2016.1238955.

Santaniello, M. (2016). Net democracy: la sfida democratica all'Internet governance. In E. De Blasio & M. Sorice (Eds.), *Innovazione democratica. Un'introduzione* (pp. 63–86). Roma: Luiss University Press.

Santaniello, M., & Palladino, N. (2017, March 30). *Shaping Words to Shape Policy Process: Discourse Coalitions in the Internet Governance Ecosystem.* Paper presented at the 1st GIG-ARTS Conference, Paris.

Schmidt, V. (2008). Discursive Institutionalism: The Explanatory Power of Ideas and Discourse. *Annual Review of Political Sciences, 11,* 303–326.

Schmidt, V. (2010). *Democracy and Legitimacy in the European Union Revisited: Output, Input and Throughput* (KFG Working Paper Series No. 21).

Take, I. (2012). Regulating the Internet Infrastructure: A Comparative Appraisal of the Legitimacy of ICANN, ITU, and the WSIS. *Regulation and Governance.* https://doi.org/10.1111/j.1748-5991.2012.01151.x.

Tallberg, J., Bäckstrand, K., & Scholte, J. A. (2018). *Legitimacy in Global Governance.* Oxford: Oxford University Press.

Van Eeten, M. J. G., & Mueller, M. (2012). Where Is the Governance in Internet Governance? *New Media & Society, 15*(5), 720–736.

Weber, R. H. (2014). Visions of Political Power: Treaty Making and Multistakeholder Understanding. In R. Radu, J. M. Chenou, & R. Weber (Eds.), *The Evolution of Global Internet Governance* (pp. 95–114). Berlin: Springer.

Weber, R. H., & Gunnarson, S. R. (2012). A Constitutional Solution for Internet Governance. *Columbia Science & Technology Law Review, 14,* 1–71.

Weinberg, J. (2000). ICANN and the Problem of Legitimacy. *Duke Law Journal, 50*(1), 187–260.

Weinberg, J. (2011). Governments, Privatization and 'Privatization': ICANN and the GAC. *Michigan Telecommunications and Technology Law Review, 18*(1), 189–218.

WGIG. (2005). *Report of the Working Group on Internet Governance.* https://www.wgig.org/docs/WGIGREPORT.pdf. Accessed 12 October 2019.

WSIS. (2005). *Tunis Agenda For the Information Society.* https://www.itu.int/net/wsis/docs2/tunis/off/6rev1.html. Accessed 12 October 2019.

Zingales, N., & Radu, R. (2015, November 9). *In Search of the Holy Grail: A Principled Approach to Multistakeholder Governance in Internet Policy-Making.* GigaNet Annual Symposium 2015. Available at SSRN: https://ssrn.com/abstract=2809920 or http://dx.doi.org/10.2139/ssrn.2809920. Accessed 12 October 2019.

Zittrain, J. (2008). *The Future of the Internet—And How to Stop It.* New Heaven and London: Yale University Press.

# CHAPTER 2

# Foundations, Pitfalls, and Assessment of Multistakeholder Governance

**Abstract** This chapter investigates the concept of multistakeholderism by drawing on contributions from global and transnational governance studies, focusing on the theoretical relationship between multistakeholderism and legitimacy. The rapid spread of multistakeholderism among governance studies and practices relies on its potential to establish legitimate authority at the global level, in which inclusive deliberative processes replace the legitimacy derived from electoral mechanisms. However, the concept of multistakeholderism reveals a structural weakness in dealing with the dimension of power, leading to governance practices that undermine less well-resourced actors. In reviewing the existing literature on the categories of input, throughput, and output legitimacy, this chapter identifies a set of legitimacy standards that a multistakeholder initiative needs to satisfy to fulfill the promises of multistakeholderism and avoid being considered merely a rhetorical exercise masking practices of domination.

**Keywords** Multistakeholderism · Global governance · Internet governance · Legitimacy

© The Author(s) 2021
N. Palladino and M. Santaniello, *Legitimacy, Power, and Inequalities in the Multistakeholder Internet Governance*,
Information Technology and Global Governance,
https://doi.org/10.1007/978-3-030-56131-4_2

## 2.1    Establishing Legitimate Authority for Transnational Governance: The Promise of Multistakeholderism

The multistakeholder concept has become an unavoidable point of reference for scholars and practitioners concerned with the institutions and decision-making processes of Internet governance. However, the concept is still under-theorized and its adoption in the Internet governance field has been largely disconnected from discussions in the broader field of global governance studies (with some notable exceptions, see, e.g., Hofmann 2016; Malcolm 2008; Raymond and DeNardis 2015; Mueller 2010; Take 2012). Accounts of multistakeholder governance in the Internet governance field often rely on a poorly elaborated theory of multistakeholderism, leading in many cases to uncritical support for the multistakeholder approach. For many observers, multistakeholderism has become "a mantra" (Hofmann 2016), "a profession of faith" (Martens 2007), and a "value in itself" (Raymond and DeNardis 2015), inhibiting debates about the possibilities of improving and changing the governance arrangements in this field.

Referring to contributions from global and transnational governance studies seems a necessary step to retrace potentiality and limitations of the multistakeholder model of governance. The concept spread in the context of global and transnational governance to describe the novelty of UN summits and initiatives open to non-state actors (such as the 1992 United Nations Conference on Environment and Development, also known as the Rio de Janeiro Earth Summit, and the 2002 Commission on Sustainable Development and World Summit on Sustainable Development held in Johannesburg), as well as new global organizations composed by a variety of non-state actors, such as the Forest Stewardship Council (1993), the Global Reporting Initiative (1997), and the World Commission on Dams (1997).

Unsurprisingly, the rise of multistakeholderism has occurred alongside the globalization of society. In this section, we point out how the proliferation of multistakeholder initiatives is strictly related to the problem of the legitimacy of international and transnational governance.

In recent decades, the volume of interdependent cross-border flows of trade, finance, and services has grown continuously. Consequently, an increasing number of common goods are provided at the transnational level, and challenges and threats tend to gain a global dimension, with the issue of climate change as the clearest example.

Emerging "new modes of governance" (Héritier and Rhodes 2011; Bäckstrand et al. 2010; Koenig-Archibugi and Zürn 2006; Risse 2006) try to address challenges that intergovernmental politics seem unable to face alone, in which traditional hierarchical methods of control centered on nation-states have been replaced by more horizontal or networked forms of coordination including non-state actors (Grewal 2008). These forms of "governance without government" (Rosenau and Czempiel 1992) or "governance with government" (Börzel and Risse 2010; Risse 2011), "government as networks" (Goldsmith and Eggers 2004), "global public policy network" (Reinicke 2000), or even "transnational regime complex" (Abbott 2012), to quote some of the most well-known attempts to conceptualize the phenomenon, have given rise to the need for descriptive and normative analysis.

In this context, the concept of global governance has moved "from the ranks of the unknown to one of the central orienting themes in the practice and study of international affairs of the post-Cold War period" (Barnett and Duvall 2005: 1) due to its ability to provide a middle ground between anarchy and the discredited idea of "world government." The global governance framework makes it possible to analyze how order can be created and rule compliance ensured at the international level, in the absence of a central political authority (Weiss 2013; Risse 2006; Buchanan and Keohane 2006).

Following Hurd (1999), rule compliance can rely on three different bases: coercion, incentives, and legitimacy. Coercion is based on the fear of punishment or sanction inflicted by rule enforcers. Rule compliance based on coercion is weak in intergovernmental governance and can be considered absent within transnational governance (Beisheim and Ding-werth 2008). Rule compliance based on self-interest entails manipulating the cost-benefit calculation of actors through incentives. Since very often global governance deals with the provision of common goods whose enjoyment is not restricted to those who produced them (e.g., pollution reduction, forest-preservation), the classic free-rider problem could undermine the incentives to bear the cost of compliance in many cases (Barrett 2007).

Because of this, an increasing number of global governance scholars have been paying attention to the concept of legitimacy. Legitimacy "facilitates compliance [...] Legitimacy is driven by the logic of appropri-ateness, whereby compliance can result from a self-imposed obligation to do what is perceived as right" (Weiss 2013: 110). Moreover, "legitimacy involves a reservoir of confidence that is not dependent on short-term satisfaction with the distributional outcomes of a given regime. Legitimacy presumes that audiences would regard an institution's exercise of authority as appropriate even if the organization were to take a decision that goes against their narrow self-interest" (Tallberg et al. 2018: 9).

Nevertheless, even the concept of legitimacy poses difficulties concerning the foundation of authority in the global sphere. In modern culture, the benchmark of legitimacy is democratic legitimacy (Bernstein 2005; Nanz and Steffel 2004), meaning that a rule is believed to be legitimate if those affected by it have participated somewhat in the decision-making process. Within contemporary domestic democracies, this participation requires representation and accountability by regular elections. Therefore, the discussion risks going down a blind alley, coming back to the starting point of the absence of a world government elected by a global political community.

It can be assumed that the "increased attention to legitimacy in global governance research is a sound scholarly reaction to intractable difficulties around the concept of global democracy" (Agné 2018: 23). Indeed, scholars and practitioners have attempted to avoid the trap related to the concept of democratic legitimacy at the global level by resorting to two different strategies.

On the one hand, they have sought to highlight effective alternative forms of legitimacy, viable for governance beyond the state. Scharpf (1999) indicated problem-solving capacity as a means to compensate for the democratic deficit of EU institutions. Suchman (1995) referred to a "pragmatic" (based on self-interest), "moral" (based on a positive evaluation of the goals and correctness of an organization), and a "cognitive" (based on the perception that the organization is necessary and inevitable) legitimacy. Other authors identified a further source of legitimacy in the expertise of decision-makers (Majone 1998; Take 2012). Recently, *Legitimacy in Global Governance* (Tallberg et al. 2018) proposed a wide range of different sources of legitimacy, including psychological and structural factors.

On the other hand, scholars have dedicated notable efforts in attempts to rethink and adapt the core concepts of democratic theory, such as participation, representation, and accountability, in a non-state-centric fashion, overcoming the mechanisms and institutions of national democracies (Bäckstrand 2006a, b; Beisheim and Dingwerth 2008; Risse 2006; Dinghwerth 2007; Macdonald 2008).

This is exactly what multistakeholderism promises to do. The success and proliferation of multistakeholder initiatives within the global governance field, and then in the sub-field of Internet governance, can be traced back to its ability to offer a viable solution for the challenges of global governance. Multistakeholderism provides a governance model capable of establishing some form of authority in the transnational sphere and claims to do so through practices embedding the core values and principles of democratic discourse (participation, representativeness, equality, and accountability), thus satisfying the political aspiration of "democratizing" global governance (Macdonald 2008; Bäckstrand 2006a).

A multistakeholder initiative can be defined as the attempt to "enable the empowered and active participation of stakeholders in the search for solutions to a common problem" (Faysse 2006: 220), where the term "stakeholder" refers to "those who have an interest in a particular decision, either as individuals or representatives of a group. This includes people who influence a decision, or can influence it, as well as those affected by it" (Hemmati 2002: 2). Practically, it usually consists of the establishment of "voluntary cooperative arrangements between actors from the public, business and civil society that display minimal degree of institutionalization, have common non-hierarchical decision-making structures and address public policy issues" (Steets 2004: 25). It is worth noting that what distinguishes multistakeholder initiatives from other governance arrangements is "their ability to bring actors from diverse backgrounds together—actors that before often had been working against each other" (Witte et al. 2000: 179), allowing a consensual solution to emerge from the combination of different views and resources.

In putting together a pluri-partite set of actors in constructive dialogue and cooperation, multistakeholderism aims to provide global politics with a deliberative model of democracy (Dryzek 2011), shifting the emphasis from the "vote-centric" to the "talk-centric" aspect of democratic legitimacy (Chambers 2003; Leeper and Slothuus 2018). In this view, recalling a conception that could be traced back to Burke, democracy is not merely, nor primarily, the making of decisions through the aggregation of pre-existing preferences. Rather, it concerns the process of preference formation through an inclusive, informed, rational, and respectful debate (Elster 1998; Steenbergen et al. 2003; Dryzek 2011; Leeper and Slothuus 2018). In this context, "democratic legitimacy resides in the right, ability, and opportunity of those subject to a collective decision to participate in deliberation about the content of that decision" (Dryzek 2009: 1381).

Besides "democratizing" governance at the global level, multistakeholderism offers two further promises, which could be considered additional sources of legitimacy.

First, multistakeholderism promises to reach a "global view" on the matters of concern (Boström and Tamm Hallström 2013). Indeed, multistakeholder initiatives "not only combine existing knowledge from different sources and backgrounds but also create new knowledge, as consensus emerges over often contentious issues" (Witte et al. 2000: 179). In doing so, multistakeholder arrangements should spur social learning and innovative thinking, leading to a broader perspective (Fransen and Kolk 2007; Cheyns and Riisgaard 2014).

Then, as argued by Bäckstrand et al. (2010) multistakeholderism promises to resolve the dilemma between democracy and effectiveness. While in traditional politics more democracy means more bargaining among parties or states, leading to compromise and less logical solutions, multistakeholder dialogue is supposed to reach higher quality outcomes. By the means of reasoned discussion, global views on the matter, and social learning, multistakeholder initiatives should help to identify the best rational solution for pursuing a shared vision of the common good. Moreover, multistakeholderism should ensure the smooth implementation of decisions, since all the actors involved in the enactment of a given initiative participated and agreed to its making.

## 2.2   STRUCTURAL PITFALLS
## OF MULTISTAKEHOLDER GOVERNANCE

In this section, we bring up the argument that the multistakeholder model of governance contains structural limitations that make it hard to adequately address the dimension of power and, for this reason, multistakeholder initiatives can easily fall short of their promises. Indeed, multistakeholder discourse could result in misleading rhetoric that solidifies power asymmetries and masks domination, manipulation, and hegemonic practices.

As noted, the end of the Cold War led to the belief in an opportunity to organize "international politics in a more inclusive and consensual manner" (Barnett and Duvall 2005: 5) often around liberal values of democracy, peace, freedom, individualism, and the market. In this view, actors are seen as interdependent and sharing common interests, thus, they should be encouraged to achieve their goals and solve their problems through cooperation and coordination, rather than unilateralism, conflict, or coercion. The absence of central political authority at the global level appears to be not only a factual statement but also often a desirable condition and a value to safeguard.

This perspective is opposed to a realism-based approach to international relations and the centrality that realism assigns to the category of power. It is worth noting that the "realist" conception of power relies on Dahl's classical definition, according to which "A has power over B, to the extent that he can get B to do something that B would not otherwise do" (1957: 202–203). This conception of power is essentially coercive and focuses on the resources that an actor can employ to force obedience or to affect the behavior of another subject, even against its will.

Global governance scholars are conscious of the differences in resources among the actors involved in transnational governance arrangements, and, in their view, the multistakeholder approach is the correct way to neutralize them, insofar as its deliberative nature should act as a guarantee against the abuse of power. Indeed, if deliberation satisfies some procedural requirement (such as being inclusive, reasoned, consensus-oriented, impartial, and respectful) is expected that "no force except the force of the better argument is at work" (Dingwerth 2007: 25), leading to a "process of mutual reason-giving" (Gutmann and Thompson 2002: 157) in which "even self-interested speakers are forced or induced to argue in terms of the public interest" (Elster 1998: 12).

Nevertheless, empirical research has pointed out that such an "ideal deliberative procedure" (Cohen 1989) has often fallen short of its expectations. Several studies highlight how multistakeholder initiatives are in many cases dominated by experts from the Global North and the private sector, while weaker stakeholders are exposed to manipulation and control due to the lack of necessary knowledge and/or resources to adequately promote their points of view during discussions (Cheyns and Riisgaard 2014; Faysse 2006; Dentoni et al. 2018), which "formal affirmative arrangements" are unable to compensate for (Fortin 2013: 573).

Moreover, the multistakeholder model itself can be a power structure producing and reproducing asymmetries and inequalities among actors and participants. Besides the classical definition of power based on the resources that actors could exploit within a *direct* social relationship, other approaches consider the possibility that power takes place within *indirect*, mediated, and diffuse social relationships (Barnett and Duvall 2005; Westergren 2016). Here, we can distinguish two different cases. First, we can refer to "institutional power" in which an indirect power relationship involves pre-constituted social actors through "frozen configurations of privilege and bias" embedded in "institutional arrangements (such as decisional rules, formalized lines of responsibility, divisions of labor, and structures of dependence)," which can shape actors' possibility of action and choices (Barnett and Duvall 2005: 15). Second, we can talk of "discursive power" in which power works through the discursive constitution of subjects, their interests, capabilities, and power positions, as well as the definition of problems, acceptable topics, arguments, and logic of appropriateness (Westergren 2016; Barnett and Duvall 2005; Cheyns and Riisgaard 2014).

These two different conceptions of power are conflated within the perspective of discursive institutionalism (Schmidt 2008, 2010) and in the concept of discourse institutionalization. According to Vivien Schmidt, institutions are created and maintained through ideas and discourses and should be considered both constraining and enabling social constructs. Similarly, Hajer's concept of "discourse institutionalization" refers to the case in which a particular discourse starts to dominate people's conceptualization of some aspect of reality, following which it "solidif[ies] into an institution, sometimes as organizational practices, sometimes as traditional ways of reasoning" (Hajer 1993: 46). In this view, actors' behavior is affected both by the means of action that the institutional design provides to them and the system of expectations that upholds the institutional setting and shapes its working practices.

The solidification of the "multistakeholderist" discourse into concrete institutional/governance practices often gives rise to a series of inconsistencies, side effects, and opportunities for dominant and powerful actors that could largely prevent multistakeholder initiatives from fulfilling their promises.

One of the most debated points concerns the consensual and deliberative approach at the base of the multistakeholder philosophy. As noted, the imperative of consensus reduces the space available for every position included within the process and may result in "depoliticization mechanisms that limit political expression and struggle" (Moog et al. 2014: 6). More radical or divergent groups may be attacked as biased or ideological and pushed to dampen their criticism and claims (Edmunds and Wollenberg 2001), "avoiding confrontation between too diverging visions" (Cheyns and Riisgaard 2014: 9). Paradoxically, subjects involved in a decision-making process because of their particular interest in the issue come to be prevented from defending their stake. In this regard, Malcolm (2008) distinguishes between consensus and deliberation, pointing out that decision-making by consensus does not necessarily require deliberation or pluralism. Thus, "consensus is the standard of agreement within many technical standards organizations" where problems could be resolved by "applying principles that are, if not entirely objective, then at least widely shared" (Malcolm 2008: 293, 311). As Malcolm notes, this condition could reflect the way of working in the first Internet community. Indeed, while consensus has been the dominant decision-making method in this field since the beginning, it does not represent the current situation.

The more this conflation of consensus and deliberation obscures the conflict among different interests and leads to the marginalization or exclusion of some points of view, the more multistakeholderism becomes a hegemonic practice and undermines its claim to provide legitimacy based on deliberative democracy. As noted by Dryzek concerning the financial governance network "if a single discourse does dominate [...] the sorts of differences and challenges that are grist for a deliberative conception of democracy get lost" and then a multistakeholder arrangement "risks becoming progressively illegitimate with time, as well as ineffective in problem-solving terms" (Dryzek 2011: 129–130).

Another point of concern relates to the role of the few self-selected actors that, despite bottom-up rhetoric, usually initiate multistakeholder processes and decide which types of stakeholder should be involved, their categorization, as well as the forms of engagement and other procedural rules (Edmunds and Wollenberg 2001; Zeyen et al. 2016; Dentoni et al. 2018; Boström and Tamm Hallström 2013). Conveners and initiators retain a great and sometimes arbitrary power that could render the process flawed from the beginning, solidifying power relationships and steering processes toward their preferred outcomes. Mueller and Wagner (2014) point out that this "bootstrapping problem," as they call it, is particularly significant in the Internet governance field due to the number of affected actors that need to be involved in the initial setup to provide the initiative with full legitimacy.

In particular, the engagement and categorization procedures of stakeholders play a crucial role. This is not only because the categorization of stakeholders is a principle of inclusion/exclusion, but also because stakeholder categories are not given or self-evident. Rather, "group formation involves complex and unstable processes of self-identification and representation" (Edmunds and Wollenberg 2001: 237). Within Internet-related governance initiatives, these dynamics have given rise to what Mueller (2010) calls "politics of representation"; an enduring struggle to set the boundaries of stakeholder groups and ensure their inclusion in the different policy arenas.

Moreover, it is worth noting that group homogeneity could affect the possibility of meaningful participation, forcing heterogeneous groups into a resource-consuming compromise or leading to the systematic marginal-ization of some elements (Tamm Hallström and Boström 2010; Fortin 2013).

Other studies (Schneiker and Joachim 2018; Seabrooke and Tsingou 2009; Tsingou 2015) have stressed the fact that selection and engagement procedures often tend to favor an already like-minded set of collective and individual actors even if they belong to different stakeholder categories. In this case, consensus "is not necessarily a product of the multistake-holder initiatives themselves [...] but rather may already exist before initiatives are established" (Schneiker and Joachim 2018: 20). Schneiker and Joachim also underline how boundaries between stakeholder cate-gories could be blurred by the so-called revolving doors phenomenon, which refers to actors who move through different organizations and stakeholder groups, raising doubts about whether they truly represent the constituencies on whose behalf they are supposed to act.

This latter aspect is particularly relevant for the Internet governance field. As noted, it is not unusual for policy-makers to wear "multiple hats" (Franklin 2013; Radu 2019; Mueller 2010), that is, they enjoy multiple affiliations. For example, a policy-maker could be a member of a tech-nical or academic community, and at the same time sit on the board of some corporation, be engaged with a governmental body, or be leading an advocacy group. While "such instances have become widely common and widely accepted in the [Internet governance] space" (Radu 2019: 180), it has been observed that multiple affiliations create confusion about which interests or constituencies a participant is acting in behalf of, raising suspicion they could serve vested interests (Franklin 2013; Mueller 2010).

Finally, it is worth noting that "paradoxically, the pressure for legiti-macy under which multi-stakeholder efforts operate creates side effects, which may ultimately undermine their legitimacy," in so far as they "tend to become increasingly complex and bureaucratic over time" (Hofmann 2016: 34). Indeed, "this increasing complexity has unintended conse-quences and counter-effects, such as making it more difficult for partic-ipants to be effective and to comprehend the entire process" (Boström and Tamm Hallström 2013: 105). Usually, complexity and bureaucracy advantage secretariats and better-resourced actors that can afford dedi-cated staff to follow expensive and time-consuming processes, with the result that "if the 'bureaucracy' becomes more and more powerful, the very multi-stakeholder feature of the multi-stakeholder organization could eventually be undermined" (ibid.).

## 2.3 A Framework to Assess the Legitimacy of Multistakeholder Initiatives

The previous sections pointed out how the success of multistakeholderism relies on the promise of establishing legitimate forms of authority at the transnational level, overcoming long-standing problems related to democratic deficit and rule compliance. However, depending on the governance arrangements and practices in which the multistakeholderist discourse takes place, it could easily be nothing more than misleading rhetoric covering and reproducing power asymmetries.

This book assesses the extent to which the multistakeholder process, named the IANA Stewardship Transition, succeeded in establishing novel legitimate governance arrangements for DNS. For this purpose, it is necessary to discuss the concept of legitimacy in greater depth and how it can be assessed.

Legitimacy is a "normative belief," which can be conceived as a "generalized perception or assumption that the actions of an entity are desirable, proper, or appropriate within some socially constructed system of norms, values, beliefs, and definitions" (Suchman 1995: 574).

As a "normative belief," the concept of legitimacy has been explored from both normative and descriptive perspectives. While the normative approach aims at identifying, by the means of philosophical-reflective methodology, the values and principles founding the "right to rule" of an entity; the descriptive approach, often referred to as the sociological approach, focuses on audiences and their reasons to believe that an authority is appropriate, desirable, and rightful, even when the authority decides against the audience's interests (Tallberg et al. 2018; Buchanan and Keohane 2006; Dingwerth 2007). The study of normative legitimacy is not confined to theorization, but meets "empirical or positive research when a scholar investigates whether a particular governance institution [...] empirically fulfills the researcher's own normatively motivated criterion" (Agné 2018: 26). Indeed, most of the empirical research on legitimacy within the context of transnational governance has been devoted to the assessment of the conformity of some transnational institution (such as the European Union, the World Trade Organization, the Forest Stewardship Council, or the Global Reporting Initiative) with some normative conception of legitimacy (Tallberg et al. 2018).

This work follows this approach, attempting to assess the extent to which the IANA Stewardship Transition process complied with normative standards of legitimacy. This choice is further justified because the NTIA announcement set some conditions for the proposal to be approved, including the adoption of a specific institutional-procedural arrangement (multistakeholder decision-making process) and the achievement of a specific objective (the transfer of the IANA functions stewardship to a multistakeholder body), which resemble normative criteria of legitimacy.

The first notable attempt to conceptualize legitimacy beyond the nation-state can be traced back to the distinction between input-oriented and output-oriented legitimacy, elaborated by Scharpf (1999) regarding the governance of the European Union. In his view, while legitimacy was grounded on different premises historically (such as religion, tradition, and charisma), in modern times, legitimacy is exclusively related or to the participatory/procedural quality of ruling institutions, or to their performative quality.

According to Scharpf, "they both start from the normative premise that legitimate government must serve the common good of the respective constituency, and that this function must be protected against both the self-interest of governors and the rent-seeking strategies of special interests" (Scharpf 2006: 2). Then, input legitimacy "would be ensured by institutions maximizing either the direct participation of the governed in policy choices or the responsiveness of governors to the (collective) preferences of the governed" (ibid.), while output legitimacy refers to the capacity of an institution to "generally represent effective solutions to common problems of the governed" (ibid.). Further, other scholars consider also a "throughput" legitimacy dimension, which refers to the "black box" of a governance system and explores the legitimacy of the process through which inputs are transformed in outputs (Schmidt 2010; Schmidt and Wood 2019).

Below, drawing on the existing literature, we trace back the normative criteria developed to assess input, throughput, and output legitimacy and define a framework to be employed in the analysis of the IANA transition process.

Concerning input legitimacy, there is a common understanding around the idea that the wide participation by the actors subjected to a decision could be a way of realizing, in place of elections, the democratic assumption according to which decision-takers will obey rules if they have participated in their making.

Unsurprisingly then, one of the most commonly mentioned criteria for assessing input legitimacy in the context of transnational governance is *inclusiveness* (Bäckstrand 2006a, b; Beisheim and Dingwerth 2008; Cheyns and Riisgaard 2014; Mena and Palazzo 2012; Shouten et al. 2012). Inclusiveness considers the extent to which all the categories of actors affected by a particular issue were involved in the decision-making process. As noted, since the categorization of actors and interests is neither given nor self-evident, evaluating inclusiveness leads to the problem of the identification, definition, and selection of relevant stakeholders and requires taking into account the implications of different alternatives (Dingwerth 2007). To operationalize the dimension of inclusiveness, Mena and Palazzo (2012) suggest considering how many different stakeholder categories were involved in a decision-making body in relation to known or declared stakeholder type, and then how many criticisms were raised due to the exclusion of some stakeholder group.

Besides the inclusion of different stakeholders, several scholars have noted that compliance with input legitimacy standards also requires their *balanced representation* (Boström and Tamm Hallström 2013; Beisheim and Dingwerth 2008; Cheyns and Riisgaard 2014). Otherwise, the underrepresentation of some category, even if included within the decision-making process, could raise the suspicion that the decision-making process has been captured by one or more special interests. There could be different criteria to balance participation. One can opt for the perfectly equal distribution of seats in a decision-making body among the identified stakeholder categories, or weigh their participation based on the size of the related constituencies (which could be particularly important in the case of geopolitical categorizations), or according to other principles. However, whichever criterion is chosen should be publicly declared and justified.

*Representativeness* constitutes another commonly highlighted principle for ensuring input legitimacy. The concept of representativeness is often used with a broad meaning that largely overlaps with the concept of inclusiveness, indicating the extent to which a decision-making body represents the political community affected by its decision.

Here, we use the term in a narrower sense, considering if the participants in a multistakeholder process are genuine representatives of the respective constituencies (Beisheim and Dingwerth 2008; Faysse 2006). In this view, representativeness relates to the accountability mechanisms within the relationship between the participants in decision-making bodies and their constituencies (Risse 2006). This requires the consideration of whether participants have been chosen and could be removed by the respective constituencies and the procedures that regulate their relationship.

In a further and less explored sense, representativeness concerns whether participants and their respective organizations represent the groups on behalf of whom they are supposed to act. It is in this context that the revolving doors phenomenon and multiple affiliations become crucial points that could undermine input legitimacy. For example, civil society associations funded or strictly connected with private firms, as well as participants belonging to different stakeholder groups, could be considered misrepresentative factors and raise the suspicion that vested interests are at work.

Throughput legitimacy relates to the process of transforming the different views, interests, and positions involved within the decision-making process into an outcome, in the form of rules, programs, guidelines, or non-binding statements. In this case, we can distinguish between legitimizing criteria related to the procedural and discursive quality of the deliberative process.

Concerning the former, many authors believe that *fairness* is a key element of procedural quality (Schmidt 2010, Beisheim and Dingwerth 2008). The assessment of the fairness of a multistakeholder initiative requires considering whether the established rules and procedures ensure that every participant has an equal opportunity to influence the process, for example, by allowing all participants to have a say in the deliberation or to vote on decisions without discrimination. Otherwise, it requires that differences among participants are morally and publicly justified, and accepted by participants (Mena and Palazzo 2012; Schouten et al. 2012; Scholte and Tallberg 2018).

Moreover, scholars frequently relate the procedural quality of a decision-making process to *accountability*, understood as a relationship in which a decision-maker is asked to report on their activities, and likely involving sanctions in the case of misconduct (Buchanan and Keohane 2006; Schmidt and Wood 2019). Global governance studies usually distinguish between internal and external accountability (Kehoane 2004; Risse 2006; Bäckstrand 2006a), where internal accountability refers to "the authorization and support which principals give to agents who are institutionally linked to one another" (Risse 2006: 185) and external accountability requires decision-makers to justify their behavior "to people or groups outside the acting entity who are nevertheless affected by it" (ibid.).

As we have seen, internal accountability could be considered a crucial component of representativeness, while external accountability is often seen as strictly related to *transparency* since transparency allows for professional, reputational, market, or funding forms of reward or sanction (Kronsel and Bäckstrand 2010; Schmidt and Wood 2019). Multistakeholder initiatives are supposed to provide information in an accessible, timely, and responsive manner regarding policy positions and the potential conflicts of interest of participants, the internal structure of responsibility, decision-making procedures, funding, and the progress of the work, allowing external stakeholders to judge the appropriateness of the process, to evaluate its activities, and to ensure that their preferences have been taken into account (Mena and Palazzo 2012; Hale 2008; Young 2001; Bäckstrand 2006a; Buchanan and Keohane 2006).

Regarding the *discursive quality* dimension, we can distinguish between two different approaches.

The first focuses on the concept of deliberativeness. It relies on deliberative democracy theory (Cohen 1989; Bächtiger et al. 2018; Chambers 1996) and, in particular, on Habermas' discourse ethics (Habermas 1999) aiming at identifying the conditions of an ideal deliberative procedure. This kind of analysis takes into account the social and communicative interactions among actors, relating democratic legitimacy to the possibility for the affected parties to take part in a decision-making process according to principles such as universality, rationality, reciprocity, impartiality, and respect. The second approach shifts the focus from the conditions of deliberation to the substantive content of the discussion. This relies on critical discursive analysis following a poststructuralist and neo-Gramscian approach (Foucault 1977; Laclau and Mouffe 1985; Howarth 2010), and in particular on the concept of *discursive democracy* (Dryzek 1990, 2011; Hajer and Wagenaar 2003) according to which "discursive legitimacy is achieved to the extent a collective decision is consistent with the constellation of discourses present in the public sphere, in the degree to which this constellation is subject to the reflective control of competent actors" (Dryzek 2011: 35).

In this view, assessing the discursive quality of a multistakeholder process means looking at the *discourse balance* (Dingwerth 2007; Schouten et al. 2012; Dryzek 2011), or in other words, it requires to take into account: (1) how many different discourses are represented within the decision-making process, compared to the different positions already existing in the issue area of concern; (2) the presence of a dominant discourse as well as of hegemonic practices aimed at inhibiting or delegitimizing different points of view; and (3) the extent to which alternative discourses have been taken into account during the discussion and in the final outcome.

Output legitimacy refers to the effectiveness of a governance arrangement in terms of problem-solving or implementation and enforcement capacity (Bäckstrand 2006a; Mena and Palazzo 2012). According to Bäckstrand (2006b), output legitimacy could be distinguished between "outcome effectiveness" and "institutional effectiveness." Outcome effectiveness refers to the problem-solving capacity of a governance system or its ability to deliver the promised results. Nevertheless, it is difficult to assess the degree to which a governance initiative reached its objectives as goals may not be sufficiently clear or could be challenging to evaluate. Also, the results may develop only in the long run, preventing any attempt to assess the efficacy of decisions until many years after they have been taken. This is true when the output of a policy process entails a concrete action aimed at solving a well-identified collective problem and is even more true in those cases, such as the one we are addressing here, that are oriented to produce new governance arrangements that cannot be considered as simple, one-dimensional outputs. As a consequence, here we use the concept of outcome effectiveness more broadly, including the long-term effects of the policy process on long-lasting conflicts among global stakeholders.

Institutional effectiveness usually takes into account if the process adopted a proper institutional design to reach its goals, usually including the establishment of a monitoring system, the degree of compliance among participants, or the enforcement capacity of the governance arrangement (Mena and Palazzo 2012; Boström and Tamm Hallström 2013). However, since the IANA transition is a constituent policy and, as such, its output produced the structural conditions affecting both the input and throughput dimensions of the novel institutional design, it is possible to assess the institutional effectiveness of the latter by analyzing it according to some criteria used to evaluate input and throughput legitimacy. While some of these criteria, such as representativeness, fairness, and discursive quality, are dependent on contingent policy processes and can be assessed only by taking into account a specific set of practices, other criteria, such as inclusiveness, balanced representation, and accountability, can be used to evaluate the structural features of the post-transition arrangement.

**Table 2.1** Types, dimensions, and criteria of multistakeholder legitimacy

| Type of legitimacy | Dimensions | Criteria |
|---|---|---|
| INPUT legitimacy | Inclusiveness | • Inclusion of all interested parties |
| | | • Criticisms for Exclusion |
| | Balanced Representation | • Proportionality |
| | | • Justification of disproportionality |
| | Representativeness | • Internal Accountability Mechanisms |
| | | • Revolving Doors |
| THROUGHPUT legitimacy | Fairness | • Equal participatory rights |
| | | • Equal participatory opportunities |
| | Accountability | • External Accountability |
| | | • Transparency |
| | Discursive Quality | • Deliberativeness |
| | | • Discourse Balance |
| OUTPUT legitimacy | Institutional Effectiveness | • Inclusiveness |
| | | • Balanced Representation |
| | | • Accountability |
| | Outcome Effectiveness | • Problem-Solving Capacity |

*Source* Authors' Creation

Table 2.1 summarizes legitimacy types, dimensions, and criteria that will be employed to assess the IANA transition process in chapters from 5 to 7.

# REFERENCES

Abbott, K. (2012). The Transnational Regime Complex for Climate Change, Environment and Planning. *Government and Policy, 30,* 571–590.

Agné, H. (2018). Legitimacy in Global Governance Research. How Normative or Sociological Should It Be? In J. Tallberg, K. Bäckstrand, & J. A. Scholte (Eds.), *Legitimacy in Global Governance: Sources, Processes, and Consequences.* Oxford: Oxford University Press.

Bächtiger, A., Dryzek, J. S., Mansbridge, J., & Warren, M. (Eds.). (2018). *The Oxford Handbook of Deliberative Democracy.* Oxford: Oxford University Press.

Bäckstrand, K. (2006a). Multi-stakeholder Partnerships for Sustainable Development: Rethinking Legitimacy, Accountability and Effectiveness. *European Environment, 16*(5), 290–306.

Bäckstrand, K. (2006b). Democratizing Global Environmental Governance? Stakeholder Democracy After the World Summit on Sustainable Development. *European Journal of International Relations, 12*(4), 467–498.

Bäckstrand, K., Khan, J., Kronsell, A., & Lövbrand, E. (2010). *Environmental Politics and Deliberative Democracy: Examining the Promise of New Modes of Governance.* Cheltenham, UK: Edward Elgar.

Barnett, M., & Duvall, R. (2005). *Power and Global Governance.* Cambridge: Cambridge University Press.

Barrett, S. (2007). *Why Cooperate? The Incentive to Supply Global Public Goods.* Oxford: Oxford University Press.

Beisheim, M., & Dingwerth, K. (2008, June). *Procedural Legitimacy and Private Transnational Governance: Are the Good Ones Doing Better?* (Report). SFB Research Center, Freie Universität Berlin, DE. https://www.sfb-governance. de/en/publikationen/sfb-700-working_papers/wp14/SFB-Governance-Wor king-Paper-14.pdf. Accessed 17 March 2020.

Bernstein, S. (2005). Legitimacy in Global Environmental Governance. *Journal of International Law and International Relations, 1*(1/2), 139–166.

Börzel, T. A., & Risse, T. (2010). Governance Without a State: Can It Work? *Regulation & Governance, 4,* 113–134.

Boström, M., & Tamm Hallström, K. (2013). Global Multi-stakeholder Standard Setters: How Fragile Are They? *Journal of Global Ethics, 9*(1), 93–110.

Buchanan, A., & Keohane, R. O. (2006). The Legitimacy of Global Governance Institutions. *Ethics & International Affairs, 20*(04), 405–437.

Chambers, S. (1996). *Reasonable Democracy: Jürgen Habermas and the Politics of Discourse.* Ithaca: Cornell University Press.

Chambers, S. (2003). Deliberative Democratic Theory. *Annual Review of Political Science, 6,* 307–326.

Cheyns, E., & Riisgaard, L. (2014). Introduction to the Symposium: The Exercise of Power Through Multistakeholder Initiatives for Sustainable Agriculture and Its Inclusion and exclusion Outcomes. *Agriculture and Human Values, 31*(3), 409–423.

Cohen, J. (1989). Deliberation and Democratic Legitimacy. In A. P. Hamlin & P. Petitt (Eds.), *The Good Polity: Normative Analysis of the State* (pp. 18–34). Oxford: Wiley.

Dahl, R. (1957). The Concept of Power. *Behavioral Science, 2*(3), 201–215.

Dentoni, D., Bitzer, V., & Schouten, G. (2018). Harnessing Wicked Problems in Multi-stakeholder Partnerships. *Journal of Business Ethics, 150*(2), 333–356.

Dingwerth, K. (2007). *The New Transnationalism, Transnational Governance and Democratic Legitimacy*. New York: Palgrave Macmillan.

Dryzek, J. S. (1990). *Discursive Democracy: Politics, Policy, and Political Science*. Cambridge: Cambridge University Press.

Dryzek, J. S. (2009). Democratization as Deliberative Capacity Building. *Comparative Political Studies, 42*(11), 1379–1402.

Dryzek, J. S. (2011). *Foundations and Frontiers of Deliberative Governance*. Oxford: Oxford University Press.

Edmunds, D., & Wollenberg, E. (2001). A Strategic Approach to Multistakeholder Negotiations. *Development and Change, 32*, 231–253.

Elster, J. (1998). Introduction. In Jon Elster (Ed.), *Deliberative Democracy* (pp. 1–18). Cambridge: Cambridge University Press.

Faysse, N. (2006). Troubles on the Way: An Analysis of the Challenges Faced by Multi-stakeholder Platforms. *Natural Resources Forum, 30*, 219–229.

Foucault, M. (1977). *Discipline and Punish*. New York: Vintage.

Fortin, E. (2013). Transnational Multi-stakeholder Sustainability Standards and Biofuels: Understanding Standards Processes. *Journal of Peasant Studies, 40*, 563–587.

Franklin, M. I. (2013). *Digital Dilemmas: Power, Resistance, and the Internet*. Oxford: Oxford University Press.

Fransen, L. W., & Kolk, A. (2007). Global Rule-Setting for Business: A Critical Analysis of Multi-stakeholder Standards. *Organization, 14*(5), 667–684.

Goldsmith, S., & Eggers, W. D. (2004). *Governing by Network: The New Shape of the Public Sector*. Washington: Brookings Institution Press.

Grewal, D. S. (2008). *Network Power: The Social Dynamics of Globalization*. New Haven, CT: Yale University Press.

Gutmann, A., & Thompson, D. F. (2002). Deliberative Democracy Beyond Process. *The Journal of Political Philosophy, 10*(2), 153–174.

Habermas, J. (1999). *Between Facts and Norms: Contributions to a Discourse Theory of Law and Democracy*. Cambridge: MIT Press.

Hajer, M. A. (1993). Discourse Coalitions and the Institutionalisation of Practise. In F. Fischer & J. Forester (Eds.), *The Argumentative Turn in Policy Analysis and Planning* (pp. 43–76). Durham and London: Duke University Press.

Hajer, M. A., & Wagenaar, H. (2003). *Deliberative Policy Analysis, Understanding Governance in the Network Society*. Cambridge: Cambridge University Press.

Hale, T. N. (2008). Transparency, Accountability and Global Governance. *Global Governance, 14,* 73–94.

Hemmati, M. (2002). *Multi-stakeholder Processes for Governance and Sustainability: Beyond Deadlock and Conflict.* London: Earthscan.

Héritier, A., & Rhodes, M. (Eds.). (2011). *New Modes of Governance in EU, Governing in the Shadow of Hierarchy.* New York, NY: Palgrave Macmillan.

Hofmann, J. (2016). Multi-stakeholderism in Internet Governance: Putting a Fiction into Practice. *Journal of Cyber Policy, 1*(1), 29–49.

Howarth, D. (2010). Power, Discourse, and Policy: Articulating a Hegemony Approach to Critical Policy Studies. *Critical Policy Studies, 3*(3–4), 309–335.

Hurd, I. (1999). Legitimacy and Authority in International Politics. *International Organization, 53*(2), 379–408.

Koenig-Archibugi, M., & Zürn, M. (2006). *New Modes of Governance in the Global System.* New York, NY: Palgrave Macmillan.

Kronsel, A., & Backstrand, K. (2010). Rationalities and Forms of Governance: A Framework for Analyzing the Legitimacy of New Modes of Governance. In K. Bäckstrand et al. (Eds.), *Environmental Politics and Deliberative Democracy, Examining the Promise of New Modes of Governance* (pp. 28–46). Cheltenham, UK: Edward Elgar.

Laclau, E., & Mouffe, C. (1985). *Hegemony and Socialist Strategy.* London: Verso.

Leeper, T., & Slothuus, R. (2018). Deliberation and Framing. In A. Bächtiger, J. S. Dryzek, & J. Mansbridge (Eds.), *Oxford Handbook of Deliberative Democracy.* Oxford: Oxford University Press. https://doi.org/10.1093/oxfordhb/9780198747369.013.37.

Macdonald, T. (2008). *Global Stakeholder Democracy, Power and Representation Beyond Liberal States.* Oxford: Oxford University Press.

Majone, G. (1998). Europe's Democratic Deficit. *European Law Journal, 4*(1), 5–28.

Malcolm, J. (2008). *Multi-stakeholder Governance and the Internet Governance Forum.* Perth: Terminus Press.

Martens, J. (2007). *Multistakeholder Partnerships. Future Models of Multilateralism? Friedrich-Ebert-Stiftung* (Occasional Paper No. 29). https://library.fes.de/pdf-files/iez/04244.pdf. Accessed 17 October 2019.

Mena, S., & Palazzo, G. (2012). Input and Output Legitimacy of Multi-stakeholder Initiatives. *Business Ethics Quarterly, 22,* 527–556.

Moog, S., Spicer, A., & Böhm, S. (2014). The Politics of Multi-stakeholder Initiatives: The Crisis of the Forest Stewardship Council. *Journal of Business Ethics.* https://doi.org/10.1007/s10551-013-2033-3.

Mueller, L. M. (2010). *Networks and States: The Global Politics of Internet Governance.* Cambridge: MIT Press.

Mueller, L. M., & Wagner, B. (2014). *Finding a Formula for Brazil: Representation and Legitimacy in Internet Governance.* Internet Policy Observatory. https://global.asc.upenn.edu/app/uploads/2014/09/Finding-a-Formula-for-Brazil-Representation-and-Legitimacy-in-Internet-Governance.pdf. Accessed 12 March 2020.

Nanz, P., & Steffel, J. (2004). Global Governance, Participation and the Public Sphere. *Government and Opposition, 39*(2), 314–335.

Radu, R. (2019). *Negotiating Internet Governance.* Oxford: Oxford University Press.

Raymond, M., & DeNardis, L. (2015). Multistakeholderism: Anatomy of an Inchoate Global Institution. *International Theory.* https://doi.org/10.1017/s1752971915000081.

Reinicke, W. H. (2000). The Other World Wide Web: Global Public Policy Networks. *Foreign Policy, 117,* 44–57.

Risse, T. (2006). Transnational Governance and Legitimacy. In A. Benz & Y. Papadopoulos (Eds.), *Governance and Democracy Comparing National, European and International Experiences* (pp. 179–199). New York: Routledge.

Risse, T. (2011). *Governance Without a State?* New York: Columbia University Press.

Rosenau, J., & Czempiel, E. O. (1992). *Governance Without Government: Order and Change in World Politics.* New York: Cambridge University Press.

Scharpf, F. W. (1999). *Governing in Europe: Effective and Democratic?* Oxford: Oxford University Press.

Scharpf, F. W. (2006). *Problem Solving Effectiveness and Democratic Accountability in the EU.* Political Science Series 107. Institute for Advanced Studies, Vienna. http://aei.pitt.edu/6097/1/pw_107.pdf. Accessed 17 November 2019.

Schmidt, V. (2008). Discursive Institutionalism: The Explanatory Power of Ideas and Discourse. *Annual Review of Political Sciences, 11,* 303–326.

Schmidt, V. (2010). *Democracy and Legitimacy in the European Union Revisited: Output, Input and Throughput* (KFG Working Paper Series No. 21).

Schmidt, V., & Wood, M. (2019). Conceptualizing Throughput Legitimacy: Procedural Mechanisms of Accountability, Transparency, Inclusiveness and Openness in EU Governance. *Public Administration.* https://doi.org/10.1111/padm.12615.

Schneiker, A., & Joachim, J. (2018). Revisiting Global Governance in Multistakeholder Initiatives: Club Governance Based on Ideational Prealignments. *Global Society, 32*(1), 2–22.

Scholte, J. A., & Tallberg, J. (2018). Theorizing the Institutional Sources of Global Governance Legitimacy. In J. Tallberg, K. Bäckstrand, & J. A. Scholte (Eds.), *Legitimacy in Global Governance: Sources, Processes, and Consequences* (pp. 56–74). Oxford: Oxford University Press.

Schouten, G., Leroy, P., & Glasbergen, P. (2012). On the Deliberative Capacity of Private Multi-stakeholder Governance: The Roundtables on Responsible Soy and Sustainable Palm Oil. *Ecological Economics, 83,* 42–50.

Suchman, M. (1995). Managing Legitimacy: Strategic and Institutional Approaches. *The Academy of Management Review, 20*(3), 571–610.

Seabrooke, L., & Tsingou, E. (2009). *Revolving Doors and Linked Ecologies in the World Economy* (CSGR Working Paper 260/09). University of Warwick, UK. http://wrap.warwick.ac.uk/1849/1/WRAP_Seabrooke_26009.pdf. Accessed 17 October 2019.

Steenbergen, M. R., Bachtiger, A., Sporndli, M., & Steiner, J. (2003). Measuring Political De-liberation: A Discourse Quality Index. *Comparative European Politics, 1,* 21–48.

Steets, J. (2004). *Developing a Framework Concept and Research Priorities for Partnership Accountability. Report for Global Public Policy Institute* (Research Paper Series No. 1).

Take, I. (2012). Regulating the Internet Infrastructure: A Comparative Appraisal of the Legitimacy of ICANN, ITU, and the WSIS. *Regulation and Governance.* https://doi.org/10.1111/j.1748-5991.2012.01151.x.

Tallberg, J., Bäckstrand, K., & Scholte, J. A. (2018). *Legitimacy in Global Governance.* Oxford: Oxford University Press.

Tamm Hallström, K., & Boström, M. (2010). *Transnational Multi-stakeholder Standardization: Organizing Fragile Non-state Authority.* Northampton, MA: Edward Elgar.

Tsingou, E. (2015). Club Governance and the Making of Global Financial Rules. *Review of International Political Economy, 22*(2), 225–256.

Weiss, T. G. (2013). *Global Governance Why What Whither.* Cambridge: Polity Press.

Westergren, M. (2016). *The Political Legitimacy of Global Governance Institutions.* Malmö: Stockholm University Press.

Witte, J., Reinicke, W. H., & Benner, T. (2000). Beyond Multila-teralism: Global Public Policy Networks. *Internationale Politik und Gesellschaft, 2,* 176–188.

Young, I. M. (2001). Activist Challenges to Deliberative Democracy. *Political Theory, 29*(5), 670–690.

Zeyen, A., Beckmann, M., & Wolters, S. (2016). Actor and Institutional Dynamics in the Development of Multistakeholder Initiatives. *Journal of Business Ethics, 135*(2), 341–360.

# CHAPTER 3

# IANA Functions, ICANN, and the DNS War

**Abstract** The chapter provides the reader with the necessary case background to understand the issues at stake within the IANA transition process. First, we describe the distributed but hierarchical architecture of the DNS, what IANA functions are, and why they are core elements of the DNS. Then we move our attention to the governance of the DNS retracing the shift from the so-called technical regime toward a "self-regulation" regime established around ICANN. In so doing, we focus on the ICANN governance structure and the controversies that undermined its legitimacy and then its authority over the DNS. Finally, we take into account the several reforms that ICANN has undertaken to comply with the multistakeholder model requirements, as well as the events, processes, and controversies that led to the IANA transition.

**Keywords** ICANN · IANA functions · Domain Name System

## 3.1 The Domain Name System and IANA Functions

The original rules defining how the Internet locates a connected resource—be it a personal computer, a mobile device, a server, a camera, a sensor, or an antenna—are specified in the Internet Protocol (IP), which

© The Author(s) 2021
N. Palladino and M. Santaniello, *Legitimacy, Power, and Inequalities in the Multistakeholder Internet Governance,*
Information Technology and Global Governance,
https://doi.org/10.1007/978-3-030-56131-4_3

is based on a system of unique addresses, each one made of a standard string of numbers. An IP address is a numerical identifier, such as 125.125.125.125, used by the network to route data traffic along different nodes from a starting point to its destination. While numerical addresses are used effectively by machines to identify and locate resources, they fall short when being managed by human users, who are better at remembering names than numbers. The need for human-readable addresses emerged among the community of the early developers of the Internet, even when there were just a few interconnected computers (*hosts*) (cfr. Liska and Stowe 2016). The tentative solution consisted of mapping names to IP addresses through a table of correspondence between the two entities. This table, the *host table*, looked like a telephone book. It was initially saved on every computer of the network and shared among the nodes in the form of a plain text file (*hosts.txt*) containing, on each row, three fields separated by white space: (i) an IP address; (ii) one or more corresponding hostnames; and iii) potentially, descriptive comments. In the beginning, the host table was manually maintained by system administrators at each node, and names were meant to be significant to the single administrator, not to the whole network. As a consequence, the same IP address could be given different names at different nodes. The need for a standardized version of the host table and unique correspondences between hostnames and IP addresses was addressed in the early 1970s, when the Network Information Center (NIC) at the Stanford Research Institute began to maintain an authoritative version of the host table on a dedicated server, the *hostname server* (Deutsch 1973; Kudlick 1974). This list, which could be occasionally downloaded by each node to get updates, developed in parallel with the list of IP addresses assigned by Jon Postel. These were registered in his *Well Known Socket Numbers* (Cerf and Postel 1972), which evolved into the *Assigned Numbers* list (Postel 1977). Both the NIC's and Postel's work were led under contracts with the Defense Advanced Research Projects Agency (DARPA), an agency of the US Department of Defense (DoD), which deals with the development of new technologies for military applications and had a fundamental role in providing the so-called founding fathers of the Internet with financial and organizational support.[1]

After a decade of exponential growth in the number of network nodes, the host table had become unmanageable for smaller hosts, and its centralized structure risked overloading the entire system. In the mid-1980s, the community of computer scientists and engineers involved in DARPA's networking projects developed a new approach to the problem of mapping hostnames to IP addresses. This new approach was based on a hierarchical and decentralized naming system, the Domain Name System (DNS). The DNS is based on the idea of subdividing the entire set of available names (the so-called domain namespace) into several zones of authority, organized hierarchically. At the top of the hierarchy is the so-called DNS root zone, consisting of two main elements: (i) the root zone file, which is the authoritative file containing the IP addresses of every subdomain and (ii) the DNS root servers, which are servers hosting the root zone file. For each subdomain, called the top-level domain (TLD), an administrator is then delegated for the maintenance of its own root zone, made up of the authoritative registry containing the IP addresses of all the names registered under the TLD as well as the TLD root server, which hosts the registry and makes it available online. At the bottom of the DNS hierarchy are the second-level domains, which are assigned from TLD registries to single domain administrators who, similarly, manage authoritative servers containing the IP addresses of every resource included within the domain, and who may autonomously subdivide their domains into further lower levels of the hierarchy. The complete domain name of a node, known as the fully qualified domain name (FQDN), comes to be an alphanumerical string composed of one or more concatenated textual labels, delimited by dots, such as www.exa mple.com. The FQDN is read from right to left, descending through the DNS hierarchy. The root is unnamed and comes after the last dot, which is usually omitted by users. In our example, "com" is the TLD under the root, "example" is the second-level domain under "com," and "www" is the third-level address specifying the exact location of the resource we are looking for under "example.com." The resolution mechanism that

---

[1] The NIC was later moved under the administrative authority of other two DoD agencies, the Defense Communications Agency (DCA), and the Defense Data Network (DDN).

allows endpoints to get the IP address of a domain name is thus made of three steps of iteration: (i) asking the DNS root server for the IP address of the "com" TLD name server, (ii) asking the "com" name server for the address of "example.com," and (iii) asking the "example.com" name server for the IP address of "www.example.com." This mechanism is implemented by DNS resolvers, which are typically located on specific servers administered by Internet service providers (ISPs).

The first set of TLDs was made of a temporary domain, ".arpa," to be added at the end of the previous names of the host table, five generic top-level domains, ".gov," ".edu," ".com," ".mil," ".org"; and several country code TLDs (ccTLDs) formed by the two-letter code identifying a country according to the ISO Standard for "Codes for the Representation of Names of Countries" (Postel and Reynolds 1984). It was decided to have thirteen DNS root servers instead of one centralized resource, with one server automatically distributing a master copy of the root file to the other twelve daily. It was also envisaged to have a set of DNS cache servers to avoid the root's congestion. In doing so, the DNS was designed as a hierarchical and distributed addressing system. The transition from the old, flat text host table to the new system started in 1983 and was completed over four years. The stewardship of this process was led by the same community of researchers who developed the system, and who started to be known as the Internet Assigned Numbers Authority (IANA). IANA is not an authority in the common sense of the word. Indeed, it was established as an informal, unincorporated, and open group of experts engaged in three main coordination activities involved in the DNS (see Internet Activities Board 1988). The first, the numbers function, consists of the allocation, registration, and maintenance of IP addresses. The second, the naming function, consists of the assignment of domain names and the management of files containing the unique correspondence between IP addresses and names. The third, the protocol parameters function, refers to the development of the standards and protocols that rule the entire addressing system of the Internet. These functions are core elements of the DNS and are referred to as "IANA functions."

## 3.2   From Technical Governance to Governance by Contract

Even though IANA people were working under contracts with the DoD, they enjoyed much autonomy over their work. IANA's functions were performed by a small, homogeneous, and well-funded community of engineers and developers (Ziewitz and Brown 2013; Hafner and Lyon 1996) who adopted an informal, pragmatic, horizontal, and open model of governance, that is commonly referred to as a "technical regime" (Hofmann 2007) or "ad hoc governance" (Castells 2001). This technical community gave rise to new institutions established through discursive interactions and mutual recognition between the involved actors, instead of formal policies and the official attribution of authority.

Within these novel decision-making arenas, personal reputation and expertise were appreciated more than institutional roles and formalized responsibilities. No formal membership or requirements were needed to join these technical bodies, whose deliberative processes were easily accessible to newcomers.

This was the case, for example, for the Internet Configuration Control Board, founded by DARPA in 1979; the Internet Activities Board (IAB), established in 1983; the Internet Engineering Task Force (IETF) in 1986, the Internet Engineering Steering Group (IESG) in 1986, and the IANA itself in 1988. Decisions over standards, technical features, and the implementation of solutions were made through informal discussions held on mailing lists and meetings, without formal voting procedures. This arrangement is referred to as "rough consensus" and is often presented together with another principle, "running code," which highlights the pragmatic orientation of decision-making toward technical issues and problem-solving (Clark 1992). In this context, an agreement was reached when a proposal met no substantial opposition, and when the proposed decision had proven to be technically feasible and effective.

The functioning of this governance model is exemplified by the Request for Comments (RFC), which consists of a technical public document used to propose Internet standards innovations and changes, and which undergoes a peer-review-like process until consensus on the proposal is reached. During this first stage of Internet development, the NIC was the official repository of RFCs (cfr. Crocker 1971) and Jon Postel was their unofficial editor (Feinler 2011).

In the early 1990s, the US government decided to gradually replace the technical regime with a more institutionalized order under the leadership of the private sector and launched a set of policy initiatives for the commercialization and privatization of the Internet, deeply transforming Internet governance arrangements (Goldsmith and Wu 2006; DeNardis and Raymond 2013; DeNardis 2014). The first move was to entrust the maintenance of the master server of the DNS root zone (server A), which had been administered by the NIC since the early 1970s, to a private for-profit company, Network Solutions Inc. (NSI), by the means of a contract to the US Defense Information Systems Agency awarded in 1991. In the same year, the invention of the World Wide Web (WWW) triggered the popularization of the Internet and turned domain names into strings with a potentially huge economic value.[2] Starting from September 1995, NSI was authorized by the National Science Foundation (NSF), which had replaced DARPA in the management of the civil network's infrastructure, to charge registration fees for second-level domain names under the TLDs.com,.net, and.org. In the same period, the first legal battles over domain names began, especially concerning trademark protection.

In 1997, the Clinton administration sanctioned the turn to DNS privatization by signing the Framework for Global Electronic Commerce, which recommended that "the private sector should lead [...] the development of a global competitive, market-based system to register Internet domain names" (Clinton 1997). The technical community tried to resist and to regain the terrain lost. The Internet Society (ISOC) had been established at the end of 1992 as an umbrella organization for the IAB, IANA, IEFT, and IESG, in a previous attempt to institutionalize the technical community's governance structure and to increase the level of its trustworthiness and legitimacy. It was equipped with professional staff and open to formal membership, both for individuals and organizations. In September 1996, ISOC launched an International Ad Hoc Committee (IAHC) to design a new international self-governance regime for the DNS root zone, based on the principles of openness and bottom-up participation developed within the IETF. ISOC involved the NSF; the World Intellectual Property Organization and the International Trademark Association, for issues concerning trademarks and copyright; and the International Telecommunications Union (ITU), the UN agency that traditionally deals with international agreements for communication networks.

---

[2] The system that was adopted by the WWW to locate web sites and pages, the Uniform Resource Location (URL), was based on domain names and used them as the prefix for content addresses, for example: www.example.com/website/webpage.html, where "/website/webpage.html" is the location of the content resource hosted on example.com (cfr. Mueller 2002: 107–108).

In December 1996, the IAHC proposed the "Generic Top-Level Domain Memorandum of Understanding" (gTLD-MoU), envisaging the establishment of a consortium based in Geneva, Switzerland, to replace NSI in the operational management of the naming functions, and prepared the launch of new TLDs to fund the organization. The US government strongly opposed this option and on January 30, 1998, the National Telecommunications and Information Administration (NTIA) issued a Green Paper, formally titled "A Proposal to Improve the Technical Management of Internet Names and Addresses,"[3] describing "the process by which the Federal government will transfer management of the Internet DNS to a private not-for-profit corporation" (NTIA 1998a). The institutional design of the new corporation was similar to that envisaged by the IAHC, the main differences being that the "new corporation" had to be established in the USA, and that "officials of governments or intergovernmental organizations should not serve on the board of the new corporation" (ibid.).

On June 5, 1998, NTIA issued its own "Statement of Policy on the Management of Internet Names and Addresses," known as the White Paper, laying the foundation for a new DNS regime (NTIA 1998b). On September 18, 1998, the Internet Corporation for Assigned Names and Numbers (ICANN) was incorporated in Los Angeles under the California Corporations Code. On November 25, 1998, a memorandum of understanding (MOU) was signed between ICANN and the US Department of Commerce (DOC). It established that ICANN, as a private sector entity, would gradually take control of the DNS and IANA functions, while the DOC would maintain "oversight of the technical management of DNS functions" (NTIA 1998c, Sec.V, Par. B, Art. 8). Further, on December 24, 1998, ICANN entered into a transition agreement with the University of Southern California to receive key personnel, computer facilities, equipment, and the intellectual property needed to perform IANA functions. On February 9, 2000, ICANN entered into an agreement with the DOC to perform IANA functions (the so-called IANA Functions Contract), which clearly stated the right of the DOC to give preliminary approval to any changes in root zone files and associated information, as well as to any modifications in equipment and personnel involved in

---

[3] The paper was published in the Federal Register on February 20, 1998 (Volume 63, Number 34).

ICANN functions. At the same time, the DOC confirmed its cooperative agreement (CA) with NSI as the maintainer of the primary root server and as the registry for the.com,.org, and.net TLDs. By doing so, the US government was able to institutionalize a separation between the power to create and assign IP numbers and domain names—a sort of legislative power that was appointed to ICANN as the IANA functions operator—and the executive power to implement those decisions on the very foundations of the Internet, which was given to NSI as the root zone maintainer. Oversight and authorization functions were retained from the US government as a consequence of NTIA's final authorization authority over changes in the root zone included both in the MOU and the CA, as well as of the American jurisdiction within which ICANN and NSI had to operate. The ICANN-NSI Registry Agreement and the ICANN-NSI Transition Agreement, both signed on November 10, 1999, come to define the relationships between ICANN and NSI. Together with the IANA Functions Contract, the separation between the IANA functions operator and the root zone maintainer came to be the main control and accountability mechanism over ICANN (Weber and Gunnarsson 2012; Post and Kehl 2015a, b). Indeed, the power to re-open IANA contract procurement to competitive bidding enabled NTIA to "extract specific, contractually-enforceable promises from ICANN concerning its governance and decision-making structure and operations" (Post and Kehl 2015a: 16).[4] Even if ICANN could manage the DNS in a private and transnational regime and the NTIA had no legal means to stop it, "without the ability to specify the contents of the Root Zone File, ICANN could no longer guarantee TLD operators that their domains would continue to exist in the DNS" (ibid.: 17).

ICANN soon entered into a complex web of contracts and agreements with other entities (for a complete overview of ICANN's contractual web, see Bygrave 2015: 50–84): with regional Internet registries (RIRs), to which IANA was used to assign blocks of IP addresses to be further delegated to local ISPs (ASO-ICANN MOU); with TLD registries that administer single TLDs (registry agreement and TLD sponsorship agreement); with registrars that sell single second-level domain names to end-users (Registration Accreditation Agreement); and with the IETF for

---

[4] The two authors report how, in 2011, NTIA effectively announced the intention to re-open the IANA Functions Contract procurement since unsatisfied by ICANN's conflict of interest policy, obtaining its prompt review by the corporation.

work related to standards for IPs (IETF-IANA MOU). At the end of the twentieth century, a private-led, contract-based regime was operating the DNS as a result of a decade of privatization.

## 3.3   The Internationalization of the DNS and the Multistakeholder Model

Through the transition agreement between ICANN and the University of Southern California, the 1998 transition was facilitated by involving the technical community and IANA people in the new corporation. However, while the technical community, after the controversy about the IAHC venture, was available to accept the new private order and to join ICANN and its constituencies, several antagonistic initiatives also emerged from the side of national governments and intergovernmental agencies. In 2003, a large group of developing countries, led by the Chinese government, asked for an international treaty for the Internet and the establishment of an intergovernmental Internet organization (Kleinwaechter 2009). Similarly, the ITU conducted intergovernmental efforts to regain authority for national governments (Kleinwaechter 2004), attempting to bring the Internet within the international institutional setting that governed transnational communication networks such as the telegraph, telephone, and satellite networks. Governments with traditionally good foreign relations with the USA, such as the European Union, Brazil, India, and South Africa, explicitly contested the new Internet order, demanding the internationalization of Internet governance, and highlighting the need to combine "the mobilisation of markets forces with the necessary leading role of governments for appropriate frameworks" (European Commission 2003: 8).[5]

Governmental concerns about the new Internet governance regime led to the organization of the World Summit on the Information Society (WSIS), which was arranged by the ITU in two phases: first in Geneva in 2003 and then in Tunis in 2005. Notwithstanding the US government's opposition to any discussion about Internet critical resources within a UN-based policy venue, the WSIS came to be focused on the DNS and the relationships between ICANN and the DoC. Because of the high level

---

[5] For a collection of other governmental reactions to the privatization of the DNS, see Mathiason (2009: 105–106).

of the controversy, at the end of the first phase, the Secretary-General of the UN set up the Working Group on Internet Governance (WGIG) to develop a working definition of Internet governance and to identify the public policy issues to be included within this policy field. The WGIG, which comprised forty members from governments, the private sector, and civil society, and which was, for this reason, the first multistakeholder body established in this area, also set out four different organizational models for the management of the DNS. Three of these proposals envisaged the replacement of ICANN with an international body, but at the end of the WSIS process, none of the proposals were accepted. ICANN remained in charge of IANA functions, and a new multistakeholder forum, the Internet Governance Forum (IGF), was established as an arena for policy debate without any binding output. The IGF was soon engaged in a battle between two factions: "forum hawks," who wanted it to become an intergovernmental mechanism for Internet governance; and "forum doves," who "emphasized those aspects of the mandate that were purely educational or informational. They were keen to prevent the IGF from becoming a starting point for disturbing the status quo" (Mueller 2010: 11). The WSIS and the IGF also became an arena for civil society representatives and the academic community to express their concerns about the private governance of Internet critical resources, as well as about the lack of an effective normative framework based on a human rights perspective (Marzouki and Jørgensen 2004; Jørgensen 2013). It is no coincidence that the first attempts to draft and formalize an Internet constitution and bill of rights, as well as "initiatives that have sought to articulate a set of political rights, governance norms, and limitations on the exercise of power on the Internet" (Redeker et al. 2018: 303), emerged soon after the establishment of the private DNS regime (Padovani and Santaniello 2018).

In June 2013 a further outbreak of contestation came following Edward Snowden's revelations about a system of mass electronic surveillance involving the US National Security Agency (NSA) and its allied governmental agencies in the so-called Five Eyes—the USA, Canada, the UK, Australia, and New Zealand. Snowden's disclosure led to worldwide condemnation of the methods used by the US government and a general sense of distrust toward the Internet's governance regime. On a policy level, the NSA leaks "landed like a bombshell into the Internet governance landscape, reigniting regime contestation and calls for the USA to relinquish its long-held special role in the key governance functions of the Internet" (Cogburn 2017: 247). Snowden's disclosures "led countries to argue that the United States was leveraging its privileged position as a hub of Internet traffic for intelligence purposes. Following these events, the number of calls to alter U.S. government involvement in the IANA functions grew" (Stifel 2017). Multistakeholderism started to lose its credibility as a viable model of Internet governance. On October 7, 2013, ICANN's president and CEO, together with CEOs of the five RIRs (AFRINIC, ARIN, APNIC, LACNIC, and RIPE NCC), the CEOs of the WWW consortium and the ISOC, and the chairs of the IAB and IETF, published a joint statement, the "Montevideo Statement on the Future of Internet Cooperation," which "expressed strong concern over the undermining of the trust and confidence of Internet users globally due to recent revelations of pervasive monitoring and surveillance" and "called for accelerating the globalization of ICANN and IANA functions, towards an environment in which all stakeholders, including all governments, participate on an equal footing" (Akplogan et al. 2013). In an attempt to regain trust and legitimacy, both at the domestic and international levels, on March 14, 2014, the US government announced its intention to transfer the stewardship of IANA functions to "the global multistakeholder community" and asked ICANN to convene global stakeholders to develop a proposal for this purpose (NTIA 2014).

## 3.4   The Evolution of ICANN Governance Structure

The first ICANN bylaw, effective as of November 6, 1998, envisaged the establishment of three supporting organizations (SOs) with the task of formulating policy proposals for the DNS:

1. The Address Supporting Organization (ASO), composed of representatives from regional Internet address registries, aimed at tackling issues concerning the operation, assignment, and management of IP addresses through an Address Council;
2. The Domain Name Supporting Organization (NSO), mainly composed of representatives from name registries and registrars of top-level domains, working on issues concerning domain names through a Names Council;

3. The Protocol Supporting Organization (PSO), composed of representatives from IP organizations that had to make proposals concerning protocol parameters, such as port numbers, enterprise numbers, and other technical parameters, through a Protocol Council.

Proposals made by SOs had to then be approved or rejected by a board of directors, which initially had nineteen members: the president *pro tempore* of the corporation, three directors nominated by each of the SOs, and nine "at-large directors" selected through an election process. The first bylaw also envisaged the creation of three advisory committees with a purely consultative role:

1. a Governmental Advisory Committee (GAC), made of representatives of national governments, multinational governmental organizations, and treaty organizations;
2. the DNS Root Server System Advisory Committee, to advise the board about the operation of the root name servers of the DNS, including security issues;
3. an Advisory Committee on Membership, which was aimed at designing a membership structure for the forthcoming election of the nine at-large directors.

The election process for the nine at-large directors proved to be controversial and ended in total failure. As reported by Klein (2004), the nine initial directors selected by the three SOs, themselves appointed "with no public participation or consultation," "in a series of board meetings in 1999 and 2000 [...] sought to eliminate, reduce, or delay electing additional directors. In so doing, they repeatedly revised the corporate bylaws that constrained board actions" (Klein 2004: 199). In the end, partial elections for just five at-large directors were held during the first eleven days of October 2000, two years after the approval of the first bylaws. Registered members could vote for one representative for their geographical region, choosing between a closed list of candidates nominated by a nominating committee (consisting of a majority of ICANN directors) and other individuals who were able to attract a minimum threshold of support via an online petition. As highlighted by Mueller (2002), the results "were stunning. In North America and Europe, the two world regions where the elections had been widely publicized and discussed, all of the candidates nominated by the ICANN's nominating committee were defeated" (Mueller 2002: 200). No other general election experiment was attempted by ICANN again. These events brought criticism against the newly institutionalized DNS regime and questioned ICANN's authority in terms of the board's accountability and member participation. Furthermore, the structure of constituencies and contracts upon which ICANN was built became unbalanced after ICANN had substantially delegated policy-making for the number and protocol functions to, respectively, the address registries and the IETF. On February 24, 2002, the then ICANN President Stuart Lynn addressed a letter "To the Internet Community," publicly highlighting server critical flaws in the ICANN organization, including the poor involvement and engagement of national governments.[6] His "case for reform" was widely discussed inside and outside ICANN and triggered the so-called Evolution and Reform Process.

The 2002 reform process resulted in the first transformation that reshaped ICANN's governance structure, the so-called new bylaws adopted on December 15, 2002. These presented four main novelties that have been retained more or less unchanged to present: the re-articulation of ICANN's constituencies, the radical change of the board's composition, new opportunities for national governments to participate through an empowered the GAC, and the introduction of some accountability mechanisms such as the Independent Review Panel (IRP) and the ombudsman.

The new bylaws changed ICANN's SOs by suppressing the PSO and replacing the NSO with two new bodies: the Generic Names Supporting Organization (GNSO), dealing with policies concerning generic TLDs; and the Country Code Names Supporting Organization (ccNSO), which was effectively established by the bylaws adopted on June 26, 2003, to manage policy processes for ccTLDs. By then, the GNSO was coordinated by a council that was structured into a bicameral house system. The Contracted Parties House hosts initiatives from ICANN's customers, namely registries and registrars, and has seven members: three appointed by the Registries Stakeholder Group (RySG), three by the Registrars Stakeholder Group (RrSG), and one by a reformed nominating

---

[6] https://archive.icann.org/en/general/lynn-reform-proposal-24feb02.htm. Accessed 12 June 2020.

committee (Nom.Com). The Non-Contracted Parties House is more heterogeneous, with thirteen members: six appointed by the Commercial Stakeholder Group (two for each of its three constituencies, the business constituency, the IP constituency, and the ISP and connectivity providers constituency), six appointed by the Non-Commercial Stakeholder Group (with its two subgroups, the Non-Commercial Users Constituency and the Not-for-Profit Operational Concerns), and one by the NomCom. The establishment of a supporting organization focused on ccTLDs gave non-American registries more visibility and power within ICANN.

As for the board's structure, voting seats became sixteen: the ICANN President ex officio; two voting members selected by each of the three SOs; one voting member selected by the newly established the At-Large Advisory Committee (ALAC), substituting for the Advisory Committee on Membership; and eight voting members selected by the NomCom. The ALAC was designed as an advisory committee representing the "interests of individual Internet users" (ICANN 2002) and currently consists of fifteen members: ten selected by five regional at-large organizations (two for each of the five regions) and five selected by the NomCom. As noted, the establishment of the ALAC resulted in a shift from a form of direct accountability through elections to a "staff-managed participatory structure that makes people part of and dependent upon the corporation" (Mueller 2009: 102).

The new NomCom currently consists of fifteen voting members: one delegated by the ASO, one by the ccNSO, one by the IAB, five by the ALAC (one for each geographical region), and seven by the GNSO, of which one is delegated by the registries, one by the registrars, one by small businesses, one by large businesses, one by ISPs, one by the IP constituency, and one by the Non-Commercial Users constituency.

Despite the reform, and frequent minor changes that occurred in the following years, ICANN has never succeeded in appeasing criticism and the claims that its structure "privileged some interests, primarily corporate and commercial" (Froomkin 2000: 71). As a consequence, "the organizational structure is under constant pressure to improve its representativeness," being "a space for various stakeholders to try to rectify what they regard as misrepresentation and imbalances of power" (Hofmann 2016: 9). The GNSO structure, in particular, has been cited as a factor producing power imbalances that penalize the Non-Commercial Stakeholder Group (Calandro et al. 2013; Mueller 2009; Gross 2011). This constituency, which represents the concerns of civil society, has suffered a condition of structural minority—overwhelmed by the convergence of the commercial interests of other stakeholder groups. Furthermore, the complex and time-consuming policy process within GNSO often prevents the full participation of this poorly resourced stakeholder group. Frequent concerns have been addressed to the ALAC regarding overlapping membership with the business sector (Gross 2011) as well as the tendency of its members to become fully integrated into the ICANN power elite structure (Mueller 2009).

Additionally, the GAC seems involved in a gradual and long-lasting process by which the advisory body has gained more power in the context of ICANN's stakeholder structure. The 2002 new bylaws gave the GAC new opportunities to participate in ICANN's policy processes, mainly through taking part in the discussion (with no voting rights) in ICANN board meetings, as well as in the process of drafting recommendations within the GNSO and the ccNSO. The new bylaws gave the GAC the possibility of adding discussion items to the board's agenda and mandated negotiations between the GAC and the board in the case that the GAC contested a board decision. According to this new provision, the board could still reject the GAC's proposals, but it had to explicitly justify its decision against conflicting advice. Another reform that expanded the GAC's power within ICANN was the Affirmation of Commitments (AoC), signed between the DOC and ICANN on September 30, 2009, which replaced the previous MOU as reworked by the 2006 Joint Project Agreement. In particular, the AoC envisaged a set of three-year review cycles on ICANN's activities and mandated that the composition of the review teams had to be agreed jointly by the chair of the GAC, the chair of the board, and some representatives of SOs. However, the reviews' outputs were meant to be non-binding on the board's decision, and, more importantly, as noted by several authors, the AoC did not affect the IANA contract which still gave "the U.S. a unilateral, life-or-death power over ICANN's authority over the DNS root zone file" (Mueller 2009; see also Froomkin 2013; Pohle and Morganti 2012).

Accountability represents another point of concern that neither the 2002 reform nor successive reform attempts have been able to tackle effectively. Here, the main problem is that the ICANN has no membership or shareholders that can constrain the ICANN board, which enjoys arbitrary decisional power over the policy development process of its community (Weber and Gunnarson 2012; Article 19 2014; Berkman Center 2010; Mueller 2009). Established appeal mechanisms, such as a request for reconsideration, the IRP, and the office of the ombudsman, have proven to be largely ineffective due to their non-binding nature (Weber and Gunnarson 2012; Article 19 2014). As observed, such shortages "casts doubt on the viability of the entire ICANN model" (Taylor 2015) and meant that "ICANN appears to be rather less accountable and democratic than its governmental counterparts" (Hofmann 2016: 39).

Notwithstanding the repeated and prolonged efforts to reform ICANN and the management of the IANA functions, three areas of contestation have persisted: the US government's oversight role, which has fed geopolitical tensions and the request to shift to an intergovernmental governance model; the corporate governance and constituency structure of ICANN, which has been criticized since the beginning for representing mostly commercial and Western interests; and the lack of accountability of the ICANN board. These issues together represent the terrain on which the request for a definitive IANA transition has grown over time.

## References

Akplogan, A. A., Curran, J., Wilson, P., Housley, R., Chehadé, F., Chair, J. A., et al. (2013). *Montevideo Statement on the Future of Internet Cooperation.* https://www.icann.org/news/announcement-2013-10-07-en. Accessed 12 June 2020.

Article 19. (2014). *ICANN Reform: Recommendations.* https://www.article19.org/data/files/medialibrary/37494/ICANN-policy-v-2.pdf. Accessed 22 January 2020.

Berkman Center. (2010). *Accountability and Transparency at ICANN. An Independent Review.* https://cyber.harvard.edu/pubrelease/icann/ Accessed 22 January 2020.

Bygrave, L. A. (2015). *Internet Governance by Contract.* Oxford: Oxford University Press.

Calandro, E., Gillwald, A., & Zingales, N. (2013). *Mapping Multistakeholderism in Internet Governance: Implications for Africa.* https://ssrn.com/abstract=2338999. Accessed 14 September 2019.

Castells, M. (2001). *The Internet Galaxy: Reflections on the Internet, Business, and Society.* Oxford: Oxford University Press.

Cerf, V., & Postel, J. (1972). *Well Known Socket Numbers, RFC322.* https://tools.ietf.org/html/rfc322. Accessed 12 June 2020.

Clark, D. D. (1992). A Cloudy Crystal Ball-Visions of the Future. In M. Davies, C. Clark, & D. Lagare (Eds.), *Proceedings of the Twenty-Fourth Internet Engineering Task Force* (pp. 540–543). Reston: Corporation for National Research Initiatives.

Clinton, W. J. (1997). *The Framework for Global Electronic Commerce.* https://clintonwhitehouse4.archives.gov/WH/New/Commerce/read.html. Accessed 12 Jun 2020.

Cogburn, D. L. (2017). *Transnational Advocacy Networks in the Information Society.* New York: Palgrave.

Crocker, S. (1971). *Distribution of NWG/RFCs Through the NIC, RFC95.* https://tools.ietf.org/html/rfc95. Accessed 12 June 2020.

DeNardis, L. (2014). *The Global War for Internet Governance.* New Haven: Yale University Press.

DeNardis, L., & Raymond, M. (2013). Thinking Clearly About Multistakeholder Internet Governance. *GigaNet: Global Internet Governance Academic Network, Annual Symposium.* http://dx.doi.org/10.2139/ssrn.2354377.

Deutsch, L. P. (1973). *Host Names On-line, RFC606.* https://tools.ietf.org/html/rfc606. Accessed 12 June 2020.

European Commission. (2003). *Towards A Global Partnership in the Information Society: EU perspective in the Context of the United Nations World Summit on the Information Society (WSIS).* COM/2003/0271 final https://eur-lex.europa.eu/legal-content/EN/TXT/HTML/?uri=CELEX:52003DC0271&from=FR. Accessed 12 June 2020.

Feinler, E. (2011). Host Tables, Top-Level Domain Names, and the Origin of Dot Com. *IEEE Annals of the History of Computing, 33*(3), 74–79.

Froomkin, A. M. (2000). Wrong Turn in Cyberspace. *Duke Law Journal, 50*(17), 17–184.

Froomkin, A. M. (2013). ICANN and the Domain Name System After the 'Affirmation of Commitments'. In Brown (Ed.), *Research Handbook on Governance of the Internet* (pp. 27–51). Cheltenham: Edward Elgar Publishing.

Goldsmith, J., & Wu, T. (2006). *Who Controls the Internet? Illusions of a Borderless World.* Oxford: Oxford University Press.

Gross, R. (2011). *Civil Society Involvement in ICANN. Strengthening Future Civil Society Influence in ICANN Policymaking.* Report. Association for Progressive Communication. https://www.apc.org/en/pubs/issue/governance/civil-society-involvement-icann-strengthening-futu. Accessed 22 January 2020.

Hafner, K., & Lyon, M. (1996). *Where Wizards Stay Up Late: The Origins of the Internet.* New York: Touchstone Books.

Hofmann, J. (2007). Internet Governance: A Regulative Idea in Flux. In R. K. J. Bandamutha (Ed.), *Internet Governance: An Introduction* (pp. 74–108). Icfai: University Press.

Hofmann, J. (2016). Multi-Stakeholderism in Internet Governance: Putting a Fiction into Practice. *Journal of Cyber Policy, 1*(1), 29–49.

ICANN. (2002). *Bylaws for Internet Corporation for Assigned Names and Numbers*. https://www.icann.org/resources/unthemed-pages/bylaws-2002-12-15-en. Accessed 17 March 2020.

Internet Activities Board. (1988). *IAB Official Protocol Standards, RFC1083*. https://tools.ietf.org/html/rfc1083. Accessed 12 June 2020.

Jørgensen, R. F. (2013). An Internet Bill of Rights? In I. Brown (Ed.), *Research Handbook on Governance of the Internet* (pp. 353–372). Cheltenham: Edward Elgar Publishing.

Klein, H. (2004). Private Governance for Global Communications. In Braman, S. (Ed.), *The Emergent Global Information Policy Regime: Technology, Contracts, and the Internet* (pp. 179–202). New York: Palgrave Macmillan.

Kleinwaechter, W. (2004). Beyond ICANN vs ITU? *The International Communication Gazette, 66*(3–4), 233–251. https://doi.org/10.1177/001654920 4043609.

Kleinwaechter, W. (2009, October 20). The History of Internet Governance. *Internet Governance*. https://web.archive.org/web/20110717120454/; http://www.intgov.net/papers/35. Accessed 12 June 2020.

Kudlick, M. D. (1974). *Host Names On-line, RFC608*. https://tools.ietf.org/html/rfc608. Accessed 12 Jun 2020.

Liska, A., & Stowe, G. (2016). *DNS Security: Defending the Domain Name System*. Cambridge, MA: Elsevier.

Marzouki, M., & Jørgensen, R. F. (2004). A Human Rights Assessment of the World Summit on the Information Society. *Information Technologies & International Development, 1*(3–4), 86–88.

Mathiason, J. (2009). *Internet Governance: The New Frontier of Global Institutions*. Abingdon, Oxon: Routledge.

Mueller, M. (2002). *Ruling the Root: Internet Governance and the Taming of Cyberspace*. Cambridge: MIT Press.

Mueller, M. (2009). ICANN Inc.: Accountability and Participation in the Governance of Critical Internet Resources. *The Korean Journal of Policy Studies, 24*(2), 91–116.

Mueller, M. (2010). *Networks and States: The Global Politics of Internet Governance*. Cambridge, MA: The MIT Press.

NTIA. (1998a, February 20). A Proposal to Improve the Technical Management of Internet Names and Addresses. *Federal Register, 63* (34), 8825–8833.

NTIA. (1998b). *Statement of Policy on the Management of Internet Names and Addresses*. https://www.ntia.doc.gov/federal-register-notice/1998/statement-policy-management-internet-names-and-addresses. Accessed 12 June 2020.

NTIA. (1998c). *Memorandum of Understanding Between the U.S. Department of Commerce and the Internet Corporation for Assigned Names and Numbers.* https://www.ntia.doc.gov/page/1998/memorandum-understanding-bet ween-us-department-commerce-and-internet-corporation-assigned-. Accessed 12 June 2020.

NTIA. (2014). *Intent to Transition Key Internet Domain Name Functions.* https://www.ntia.doc.gov/press-release/2014/ntia-announces-intent-transition-key-internet-domain-name-functions. Accessed 12 June 2020.

Padovani, C., & Santaniello, M. (2018). Digital Constitutionalism: Human Rights and Power Limitation in the Internet Eco-system. *The International Communication Gazette, 80*(4), 295–301. https://doi.org/10.1177/174804 8518757114.

Pohle, J., & Morganti, L. (2012). The Internet Corporation for Assigned Names and Numbers (ICANN): Origins, Stakes and Tensions. *Revue Française D'études Américaines, 4*(134), 29–46.

Post, D. G., & Kehl, D. (2015a). *Controlling Internet Infrastructure: The 'IANA Transition' and Why It Matters for the Future of the Internet, Part 1.* https://static.newamerica.org/attachments/2964-controlling-internet-infras tructure/IANA_Paper_No_1_Final.32d31198a3da4e0d859f989306f6d480. pdf. Accessed 12 June 2020.

Post, D. G., & Kehl, D. (2015b). *Controlling Internet Infrastructure: The 'IANA Transition' and ICANN Accountability, Part 2.* https://static. newamerica.org/attachments/9764-controlling-internet-infrastructure-2/ IANA_Paper_2_final.8594b4de27dd4ecf9be46d348f848cf1.pdf. Accessed 12 June 2020.

Postel, J. (1977). *Assigned Numbers, RFC739.* https://tools.ietf.org/html/ rfc739. Accessed 12 June 2020.

Postel, J., & Reynolds, J. (1984). *Domain Requirements, RFC920.* https://tools. ietf.org/html/rfc920. Accessed 12 June 2020.

Redeker, D., Gill, L., & Gasser, U. (2018). Towards Digital Constitutionalism? Mapping Attempts to Craft an Internet Bill of Rights. *The International Communication Gazette, 80*(4), 302–319. https://doi.org/10.1177/174804 8518757121.

Stifel, M. (2017). *Maintaining U.S. Leadership on Internet Governance. Council on Foreign Relations.* https://www.cfr.org/report/maintaining-us-leadership-internet-governance. Accessed 12 June 2020.

Taylor, E. (2015, September 21). The Internet Is Run by an Unaccountable Private Company. This Is a Problem. *The Guardian.* http://www.the guardian.com/technology/2015/sep/21/icann-internet-us-government. Accessed 22 January 2020.

Weber, R. H., & Gunnarson, S. R. (2012). A Constitutional Solution for Internet Governance. *Columbia Science & Technology Law Review, 14,* 1–71.

Ziewitz, M., & Brown, I. (2013). A Prehistory of Internet Governance. In I. Brown (Ed.), *Research Handbook on Governance of the Internet* (pp. 3–26). Cheltenham: Edward Elgar Publishing.

CHAPTER 4

# The Institutional Design of the IANA Transition Process

**Abstract** This chapter provides readers with a detailed overview of the IANA transition process, tracing back the steps that have led to the drafting of the transition proposal. The chapter introduces the procedural elements of the initiative describing the composition, rules of engagement, and approval procedure of the different bodies entrusted to draft the proposal, and draws on the huge amount of available documentation, including official statements, charters, public comments and submissions, meeting transcripts, and reports. It also gives an account of the content of deliberation in different sites and how it changed during the process.

**Keywords** ICANN · IANA transition · Institutional design

The IANA transition process has been the result of more than two years of intense work and discussions that, according to the official data released by ICANN, has taken more than 800 working hours, 600 calls and meetings, and 33,100 mailing list exchanges.[1] The process has been extremely complex, involving six main working/coordination groups in

---

[1] https://www.icann.org/resources/pages/iana-accountability-participation-statistics-2015-11-04-en. All the online resources mentioned in this chapter were last accessed on December 2019.

© The Author(s) 2021                                                                   63
N. Palladino and M. Santaniello, *Legitimacy, Power, and Inequalities in the Multistakeholder Internet Governance,*
Information Technology and Global Governance,
https://doi.org/10.1007/978-3-030-56131-4_4

turn divided into a myriad of sub-groups entrusted with particular tasks. This complexity makes it difficult for external observers to understand how the IANA transition proposal has been developed. For this reason, this chapter serves mainly a descriptive purpose to the reader with an as complete as possible picture of the process, tracing back all the steps that have led to the drafting of the transition proposal.

In so doing, we pay attention to the procedural aspects highlighted in Chapter 2 that are relevant to assessing the legitimacy of the multistakeholder process. We take into account the charters, composition, and rules of engagement of the different bodies entrusted to draft the proposal.

## 4.1   THE PREPARATORY PHASE

With the announcement of the IANA transition process, the US National Telecommunications and Information Administration (NTIA) established the following conditions that the submitted proposal should meet to be approved:

1. Support and enhance the multistakeholder model; meaning that "NTIA will not accept a proposal that replaces the NTIA role with a government-led or an inter-governmental organization solution"[2];
2. Maintain the security, stability, and resiliency of the Internet DNS;
3. Meet the needs and expectations of the global customers and partners of the IANA services;
4. Maintain the openness of the Internet.

In the same document, the NTIA entrusted ICANN with the task of convening a "multistakeholder process to develop the transition plan," arguing that as the "current IANA functions contractor and the global coordinator for the DNS," ICANN was the best-positioned actor to carry out this duty. Yet, the NTIA requested that ICANN "work collaboratively with the directly affected parties, including the Internet Engineering Task Force (IETF), the Internet Architecture Board (IAB), the Internet Society (ISOC), the Regional Internet Registries (RIRs), top-level domain name operators, VeriSign, and other interested global

---

[2] https://www.ntia.doc.gov/press-release/2014/ntia-announces-intent-transition-key-internet-domain-name-functions. Accessed 12 June 2020.

stakeholders." ICANN discussed the design of the mechanisms for the transition during the March 24 session at its 49th meeting in Singapore in 2014. Besides the people physically in Singapore, the session was open to remote participation through a mailing list (ianatransition@icann.org).

On April 8, ICANN posted a "Call for Public Input: Draft Proposal, Based on Initial Community Feedback, of the Principles and Mechanisms and the Process to Develop a Proposal to Transition NTIA's Stewardship of the IANA Functions" on its website,[3] setting out a framework for the transition and asking for further comments. ICANN also issued a "Scoping Document"[4] delimiting the goals and objectives of the transition and the topics that should have been included or excluded by the process. This scoping document clarified that the transition proposal had to "focus on defining accountability mechanisms that would serve to replace the current stewardship role played by NTIA to ensure ICANN's performance of the IANA functions." Further, the structure of the policy development process related to the IANA functions, the role of ICANN as the IANA function operator, as well as issues not related to the IANA functions (e.g., cybersecurity, privacy, and content), were considered outside the scope of the proposal.

The call for input added that the issue of improving the globalization and accountability of ICANN should not be addressed by the stewardship transition process, although it was recognized as strictly interrelated with ICANN's role as the IANA functions operator. Instead, the call proposed to carry out these issues through a parallel but separated process, all internal to the ICANN community (its supporting organizations and advisory committees).

Further, the ICANN board proposed the establishment of a steering group to coordinate the process and ensure its compliance with the established timelines, principles, and requirements. Two representatives from each supporting organization/advisory committee group,[5] selected by

---

[3] https://www.icann.org/resources/pages/draft-proposal-2014-04-08-en. Accessed 12 June 2020.

[4] https://www.icann.org/en/system/files/files/iana-transition-scoping-08apr14-en.pdf. Accessed 12 June 2020.

[5] As discussed in Chapter 3, ICANN's supporting organizarions (SOs) and advisory committees (ACs) are the Generic Names Supporting Organization (GNSO); the Country Code Names Supporting Organization (ccNSO); the Address Supporting Organization (ASO). ICANN has the following Advisory Committees: Governmental Advisory

the chair of the ICANN board and the chair of the Governmental Advisory Committee (GAC), were to comprise the steering group, together with two representatives selected by each of the affected parties (the IETF, the IAB, the ISOC, and the RIRs). Being the convener, the ICANN board would have appointed one participant as a board liaison to the steering group and provided it with an ICANN secretariat.

This first proposal by the ICANN board was met with strong and widespread opposition among the Internet community. Several comments pointed out that ICANN, in its dual role as the convener of the IANA transition process and the IANA functions operator, was in a situation of potential conflict of interest.[6] In their opinion, the process designed in the call for input could arouse suspicion about ICANN's intention of taking advantage of its position to solidify its control over IANA functions.

Two points, more than others, raised concerns: (1) the "ICANN-centrism" of the designed process, which was entirely based on ICANN structures and with members chosen by the ICANN board; and (2) the delimitation of the scope of the transition beyond NTIA requirements, which was considered an unacceptable limitation of the discussion. Consequently, they demanded a process that went beyond the ICANN ecosystem and claimed that it should be up to the community itself to define both the composition of the coordination group and the scope of the process. Some other comments suggested removing the ICANN framework altogether and establishing an "ad hoc" multistakeholder initiative with a more inclusive structure.

On June 8, the ICANN board released the "Process to Develop the Proposal and Next Steps," which did not take into account alternative frameworks for the transition and was mainly focused on the establishment of the steering group, renamed the "coordination group" to dispel concerns about it taking a top-down approach. However, even in this renewed process, the coordination group composition and members' selection relied on the ICANN community and its affected parties. Nevertheless, the board brought several changes to meet some of the requests raised in the public comments:

Committee (GAC); Security and Stability Advisory Committee (SSAC); Root Server System Advisory Committee (RSSAC); At-Large Advisory Committee (ALAC).

[6] Comments are available at: http://mm.icann.org/pipermail/ianatransition/2014/date.html#0. Accessed 12 June 2020.

1. The members of the coordination group would no longer be appointed by the chairs of the ICANN board and the GAC; rather, each community would select its representatives. Moreover, the board asked the communities to take into account geography and the home country development level of their representative during the selection process, according to the "diversity" principle. However, no formal mechanisms were put in place to check compliance with this criterion.
2. The composition of the steering/coordination group was changed, providing additional seats for gTLD operators and ccTLD registries not belonging to ccNSO. An additional seat was reserved for the International Chamber of Commerce Business Action to Support the Information Society (ICC-BASIS) to include a broader cross-section of business representation. The number of GNSO representatives increased from two to three (selected from the non-registry component), while the number of address registry appointees was lowered, following the numerous complaints about the overrepresentation of this category. After the establishment of the coordination group, the GAC obtained another three seats.
3. To dampen suspicions of ICANN's non-neutral role, the new scheme established an independent secretariat providing the coordination group with administrative and logistical support. The secretariat would have been funded by ICANN but selected by the coordination group itself through an open process.
4. The document left the precise definition of the coordination group's roles to the input provided by the global multistakeholder community.

It is worth noting that the document directly launched the call for the creation of the coordination group asking the stakeholders to select their representatives, without having before voted on the proposed scheme, neither within the ICANN community nor outside through a further round of public comments.

In the meantime, an "enhancing ICANN accountability" process was launched. Following the Singapore meeting and the first exchange of comments and opinions, on May 6, 2014, the ICANN board released its first proposal to identify how ICANN could remain accountable in the absence of its historical relationship with the US government.[7] For this purpose, the ICANN board tabled the establishment of a working group, whose members would be appointed in an equal number by ICANN supporting organizations and advisory committees. ICANN staff would have selected external experts in technical, governance, economic, or juridical matters to join the working group.

A discussion followed, including the first round of comments, the reply of the ICANN board, and a new period of public comments, which pointed out how the interested stakeholders deemed the proposal incomplete and controversial.[8] The most debated points concerned:

1. The composition of the working group: most of the comments agreed that the community was supposed to decide the composition of the working group, defining it as the broader Internet community, not just the ICANN community. Moreover, the leadership of ICANN community supporting organizations, advisory committees, stakeholder groups, and constituencies issued a joint statement expressing their preference for the Cross-Community Working Group (CCWG) model, an already existing and tested framework within ICANN's governance practices allowing broad participation open to everyone interested, except for consensus calls which were reserved to designated supporting organization/advisory committee members.[9] In its communication of October 10, 2014, ICANN embraced the CCWG model.[10]

2. The role and selection of experts: many commenters criticized the appointment of experts by the ICANN board, claiming that it would have meant a conflict of interest and a violation of ICANN's supposed neutral role. Instead, they proposed that the community should have selected the experts based on their needs, and advisors would not have voting rights within consensus calls. In the end, ICANN decided to establish a "public experts" group, comprised

[7] https://www.icann.org/resources/pages/enhancing-accountability-2014-05-06-en. Accessed 12 June 2020.

[8] All the mentioned material can be found at: https://community.icann.org/display/acctcrosscomm/General+Process+Information. Accessed 12 June 2020.

[9] https://forum.icann.org/lists/comments-enhancing-accountability-06sep14/pdfAQjmNDuA3i.pdf. Accessed 12 June 2020.

[10] https://www.icann.org/resources/pages/process-next-steps-2014-10-10-en. Accessed 12 June 2020.

of four external and independent experts selected by the board itself and charged with the identification of seven advisors, without voting power, to be included in the CCWG.

3. Charter and scope: the whole ICANN community, as represented by the joint statement, together with external stakeholders, rejected the board's intention to draft the charter of the working group and claimed that the task was the responsibility of to the community itself. In particular, comments contested the narrowness of the scope of the process established by ICANN, which was limited to the identification of a new mechanism to replace the NTIA's oversight. Many contributors instead wanted the transition as an opportunity to address and resolve the long-standing dissatisfaction with the available instruments to review and influence ICANN board decisions. Many comments maintained that ICANN's position on chartering and scope had revealed a conflict of interest that threatened to undermine the overall process. In the end, the community was allowed to draft the charter, including the scope of the process.

4. Timeline and relation with the IANA transition: some contributions raised concerns about the decision to divide accountability in the transition process as they are strictly interrelated, and stressed the idea that their separation could have undermined the consistency and efficacy of the overall initiative. In the face of the board's refusal to reconsider this point, most of the comments pointed out that the accountability process had to be closed before the transition would take place. To avoid contradicting the wide consensus around a broad mandate for the CCWG, some contributions proposed dividing the activities of the accountability process into two different streams. Work Stream 1 focused on mechanisms enhancing ICANN accountability that had to be in place within the time frame of the IANA Stewardship Transition. Work Stream 2 would have addressed accountability topics with timelines that may have been extended beyond the IANA Stewardship Transition. The board adopted the proposal.

## 4.2   The Drafting of the IANA Stewardship Transition Proposal

In July 2014, the IANA Stewardship Transition Coordination Group (ICG) held its first meeting in London. On August 27, the ICG released a charter[11] establishing that, as previously suggested by IAB during the preparatory phase,[12] the drafting of the proposal for the IANA transition would be split into three different parts, concerning the name, number, and protocol parameter functions. Since each of these poses distinct organizational, operational, and technical issues, the ICG entrusted the "operational community" developed around each of the IANA functions to draft their respective part of the proposal.

Moreover, the charter specified the following tasks for the ICG: (i) acting as a liaison among the different operational communities, ensuring that each community was progressing in drafting its part of the transition proposal; (ii) assessing the compatibility and interoperability among proposals, as well as the consensus reached within the community; and (iii) assembling the partial drafts into a final proposal to be submitted to the NTIA.

On September 8, a request for proposal was launched, specifying that the "operational communities" of the IANA consist of "those community with direct operational or service relationships with the IANA functions operator, in connection with names, numbers, or protocol parameters."[13] The request for proposal also specified the elements that every partial proposal was requested to address: (1) a description of the community's use of IANA functions; (2) existing pre-transition arrangements; (3) proposed post-transition oversight and accountability arrangements; (4) a test of the transition implications, including potential risks, legal framework requirements, and workability; (5) consistency with NTIA requirements; and (6) a description of the community process, including the steps taken to develop the proposal and to achieve consensus.

On September 17, the ICG released its guidelines for decision-making, based on a consensual approach. According to these, the approval of the IGC's recommendations would require the full agreement of all members

---

[11] https://www.icann.org/en/system/files/files/charter-icg-27aug14-en.pdf.   Accessed 12 June 2020.

[12] https://www.iab.org/wp-content/IAB-uploads/2014/04/iab-response-to-201 40408-20140428a.pdf. Accessed 12 June 2020.

[13] https://www.icann.org/en/system/files/files/rfp-iana-stewardship-08sep14-en.pdf. Accessed 12 June 2020.

or with the motivated opposition of a small minority. The guidelines specified that the determination of what is a "small minority" would be made on a case-by-case basis by the chairs and that minority views would be documented and reported. Moreover, the guidelines clarified that the objection of the majority of an operational community would prevent the submission of a final proposal.

In December 2014, the ICG published a further document explaining the process of finalizing and assembling the proposal. For the first time, the ICG would check whether the proposal of each operational community was consensus-based, complete (concerning the components specified in the request for proposal), clear, and respectful of NTIA criteria. Then, the ICG would assess the interoperability, accountability, and workability of their combination. In case of problems, the ICG would ask the operational communities to address them, avoiding drafting a single proposal alone.

### 4.2.1  The Name Proposal

In the summer of 2014, the ccNSO, the GAC, the GNSO, the ALAC, and the SSAC established a CCWG to develop an IANA transition proposal on naming related functions (CWG-Stewardship). The CWG-Stewardship drafted a charter, which was adopted by the chartering organizations between the end of August and October 2014.

According to the charter, every chartering organization could appoint between two and five members to the working group, following their own rules and procedures. In the end, the ALAC, the ccNSO, and the GNSO appointed five members each, while the GAC and the SSAC appointed two. The chartering organizations were also entitled to appoint the CWG-Stewardship chairs, who were responsible for presiding over the group's deliberation, ensuring the bottom-up nature of the process, balanced participation, and that no one would threaten to disrupt the working group.

Moreover, the charter provided people interested in the accountability process with the possibility of joining the activities of the CWG-Stewardship as "participants," who were allowed to attend the group's meetings and to intervene during face-to-face and mailing list discussions. In the end, there were 137 listed individual participants. For the most part, they were members, or former members, of the ICANN community.

Concerning the rules of engagement, the CWG-Stewardship charter provided that decisions about the proposal, reports, and work plan were to be taken by consensus, and distinguished between "full consensus," in the absence of formal objections, or "consensus," indicating that only a small minority disagrees. In the latter case, the charter charged the chairs with the responsibility of defining when a minority is "small," and therefore when the level of consensus was to be considered satisfactory. The minority were permitted to attach minority statements to the related report. According to the charter, only the members appointed by the chartering organizations could take part in consensus calls.

Later, the CWG-Stewardship established a working sub-group for every item included in the ICG's request for proposal. The sub-groups included volunteer rapporteurs and internal coordinators and reported to the CWG-Stewardship as a whole, which discussed their deliverables and finally approved them according to the decision-making rules established in the charter.

A first draft proposal was ready on December 1, 2014.[14] At its core, the proposal suggested replicating the existing arrangements of a formal contract between the IANA function operator and an independent oversight entity. While the CWG-Stewardship agreed to not question the role of ICANN as the IANA function operator, the proposal replaced the NTIA's oversight functions with a newly formed nonprofit corporation. The so-called Contract Co. would have been independent and separate from ICANN and designed as a lightweight body, with little or no staff, whose primary function would be to enter into a contract with the IANA functions operator for the performance of the IANA functions. Contract Co. was conceived as the operative and legal arm of a "multistakeholder review team" (MRT), a multistakeholder body that had to define the terms of the IANA function contract, review the budget and the performance of the IANA function operator, and make the key decisions, including on the possible replacement of the IANA function operator. As specified at page 67 in the draft proposal "the MRT would be a multistakeholder body with seats allocated to all relevant communities."

The public comment period (December 1–22) brought to light a general agreement on the principle of (potential) separability between

[14] https://community.icann.org/download/attachments/50823498/CWG-Dec01Publ icConsultFINAL.pdf?version=1&modificationDate=1417638162000&api=v2. Accessed 12 June 2020.

ICANN and the IANA functions operator.[15] Nevertheless, a powerful minority (including among others, the Business Community, the ALAC, Google, Nominet, i2Coalition, and ICANN itself) opposed the creation of an external contracting entity, preferring solutions internal to ICANN. Moreover, several commenters raised concerns about the complexity and the lack of completeness and clarity of the proposal.

Based on this feedback, the CWG-Stewardship decided to carry forward further discussions taking into account a range of alternative solutions with the support of independent legal advice. After presenting seven different models at the Singapore meeting in February 2015, ten further workgroups, called "design teams," were established, with each focused on delivering outputs on a specific unsolved issue in a short timeframe.

During the ICANN Istanbul meeting of March 2015, attenders debated the seven models and the proposals of the design teams with the support of the selected legal counsel, Sidley Austin LLP, who exposed each model and answered the questions raised by participants. The dialogue with the legal counsel progressed beyond the Istanbul meeting and led to choosing the option of an IANA functions operator internal to ICANN, but legally separate from it.

A second draft proposal was published for public comment on April 22.[16] According to this, the NTIA would have transferred IANA functions, related resources, and the rights for contracting IANA functions directly to ICANN itself. In turn, ICANN would contract a new legal entity, the Post-Transition IANA (PTI), established as an affiliate nonprofit corporation controlled by ICANN, and grant it the rights and obligations to serve as the IANA functions operator. At the same time, a series of novel processes and bodies (such as the IANA Function Review and the Customer Standing Committee) would be established to oversee the performance of the IANA function operator and eventually begin an escalation mechanism leading to a separation process.

The draft was refined, with minor changes, into a final proposal between June 1 and 9, and delivered to the chartering organizations for their approval and then submitted to the ICG. All the chartering

---

[15] All the comments are available at: https://www.icann.org/public-comments/cwg-naming-transition-2014-12-01-en. Accessed 12 June 2020.

[16] https://community.icann.org/download/attachments/53773326/cwg-stewardship-draft-proposal-with-annexes-22apr15-en.pdf?version=1&modificationDate=143051222 3000&api=v2. Accessed 12 June 2020.

organizations supported the final proposal, pointing out that it was significantly dependent on, and expressly conditioned by, the implementation of ICANN-level accountability mechanisms by the CCWG-Accountability, whose failure in the development of satisfactory arrangements would have required the revision and a re-approval of the final proposal.

### 4.2.2    The Number and Protocol Proposal

The number and protocol communities followed a more straightforward and faster process.[17] The RIRs established the Consolidated RIR IANA Stewardship Proposal Team (CRISP Team), convening members of its regional organizations. According to its charter, each of the five RIRs could appoint two voting members and a staff member to the team. The charter specified that the CRISP Team would carry out its work through teleconferences and a public mailing list. Moreover, the CRISP Team was expected to approve its outcome via "rough consensus" or with a supermajority of eight out of ten. A first draft of the transition proposal was published on December 19, 2014, and was left open to public comment until January 5, 2015. A second draft was then released on January 8, 2015, and, after a further period of public comment, was revised and sent to the ICG on January 15, 2015.

IETF, in the behalf of the protocol community, established the IANA-PLAN working group, drawing on the usual IETF decision-making procedures. Thus, the discussion was open to anyone and decisions were to be made by "rough consensus." The working group concluded its draft proposal in November 2014. The IESG and the IAB approved and submitted the proposal to the ICG on January 6, 2015.

The proposals of the number community and those of the protocol parameter community had a similar structure and shared several key points. Both moved from the assumption that IANA customers (i.e., the number community, the protocol parameter community, and the name community) should have independent arrangements with the IANA functions operator related to the maintenance of the specific registries that

---

[17] Materials on the work of the Number Community can be found at https://www.nro.net/internet-governance/iana/iana-stewardship-transition/stewardship-transition-archive/. For the protocol community, refer to https://datatracker.ietf.org/wg/ianaplan/about/. Accessed 12 June 2020.

they are responsible for. Both the communities declared their satisfaction with the performance of ICANN as the IANA function operator and asked that ICANN maintain its role to safeguard the security, stability, and continuity of DNS operations, with only minimal changes to the existing arrangements. The number community suggested the establishment of a contractual Service Level Agreement (SLA) be established between the Regional Internet Registries and the IANA Numbering Services Operator, and a Review Committee (RC) composed by community representatives from each region to advise the RIRs to monitor IANA function operator performances. The protocol parameter community proposed to continue to rely on the current Memorandum of Understanding between IETF and ICANN.[18]

### 4.2.3 The Assembled Proposal

Following the submissions from the three operational communities, the ICG assessed the proposals and assembled them into a unique and complete document. The Combined Transition Proposal was published on July 31 for a public comment period that ended on September 8.[19] The ICG reviewed the submitted comments during a two-day face-to-face meeting (September 18–19) and selected the questions to be forwarded to the respective operational communities for further consideration.

A minority of comments touched on several substantial issues, such as jurisdiction, the lack of independence of the PTI from ICANN, and its relationship with the number and protocol communities. A broader number of comments raised concerns about the complexity of the proposal and the possible consequences for interoperability and workability, as well as about its incompleteness (especially on various aspects of the PTI), and its dependence on the work of the CCWG-Accountability. Nevertheless, the ICG found that the majority of comments were supportive of the proposal and worked with the appropriate operational community to fix the details.

[18] https://tools.ietf.org/html/rfc2860. Accessed 12 June 2020.

[19] https://www.ianacg.org/calls-for-input/combined-proposal-public-comment-period/. Accessed 12 June 2020.

On October 29, after an ICANN meeting held in Dublin, the ICG announced that it had finalized the IANA Stewardship Transition Proposal. It declared that the name proposal was conditioned on the accountability mechanisms under development in the CCWG-Accountability, which would ensure that the accountability requirements advanced by the name community would be satisfied before transmitting the final proposal to the board and the NTIA.

## 4.3    The Drafting of the Enhancing ICANN Accountability Recommendations

Following public comment periods in the preparatory phase, a drafting team comprised of representatives of ICANN's supporting organizations and advisory committees was established with the task of drafting the Charter of the Cross-Community Working Group on Enhancing ICANN Accountability (CCWG-Accountability), which was ready for circulation on November 3, 2014, and adopted by chartering organizations between November and December.

Similar to the CWG-Stewardship, the CCWG-Accountability charter provided that the chartering organizations would appoint their voting members (from two to five), the chair would allow interested people to take part in the group as non-voting participants, and would establish decision-making rules based on "full consensus" or "consensus," with the possibility of including minority statements in the reports of the groups. Different from the CWG-Stewardship charter, however, an ICANN board liaison, an ICANN staff representative, a former Accountability and Transparency Review Team[20] member, and seven advisors selected by the Public Experts Group would also be included in the CCWG-Accountability as active members of the group. However, they were not allowed to take part in consensus calls.

The charter also provided that the ICANN board could oppose the implementation of a CCWG recommendation if it determined that it was not in the "global public interest," with a two-thirds majority. In the end, the ALAC, the ccNSO, the GAC, and the GNSO appointed five members

---

[20]Voulontary working groups entrusted to conduct periodic review of ICANN's accountability and transparency.

each, the ASO appointed four, and the SSAC two. A total of 175 individual participants—for the most part members, or former members, of the ICANN community—joined the group.

The CCWG-Accountability first met in December 2014 and organized its activities across four different "work areas": (1) Inventory of Existing Accountability Mechanisms, (2) Assessment of Comments to Date, (3) Interrelation with the CWG-Stewardship Work, and (4) Stress Test and Contingencies Work Party. These groups mainly provided a preliminary assessment of existing mechanisms and defined the job to be done. At the Frankfurt meeting on January 19–20, 2015, the groups were restructured into two "work parties" to move toward the development phase of the proposal. The work parties focused on mechanisms related to the "Community Empowerment" (WP1) and "Review and Redress" of the board decision (WP2).[21] Further working groups were established to conduct "stress-tests" and to ensure the resilience of the proposed arrangements to unexpected events.

In March 2015, the CCWG-Accountability engaged two legal firms, the already-mentioned Sidley Austin LLP and Adler & Colvin, to receive advice on the feasibility of the arrangements conceived for the proposal. After two further face-to-face meetings in Singapore (February 9–12, 2015) and Istanbul (March 23–24, 2015) as well as two intense workdays (April 23–24, 2015) a first draft proposal was published on May 4, 2015, and left open to public comment until June 12.

The proposal was expressly drawn on a "state analogy" based on four "building blocks"[22]:

1. an "Empowered Community," corresponding in this metaphor to the people and meaning that the supporting organizations and advisory committees comprising the ICANN community would be endowed with the power to recall the ICANN board or some of its

---

[21] Later, new work parties were set up. WP3 "Emerging Issues" was established in July 2015 to deal with the accountability of supporting organisations, advisory committees, and ICANN staff. WP4 "Human Rights" was created in August 2015 to discuss the inclusion of a commitment to human rights in ICANN's mission and bylaws. Stress Test 18 Work Party (ST18-WP) was established in November 2015 to address changes in the bylaw concerning the GAC. Finally, the Work Party–IRP Implementation Oversight Team (WP-IOT) was convened in January 2016.

[22] See p. 14 of the First Draft Proposal available at: https://community.icann.org/pages/viewpage.action?pageId=52897394. Accessed 12 June 2020.

members, reject board decisions on the strategic plan and budget, and an opposing veto to bylaw changes;

2. a set of constitution-like principles comprising the mission, commitments, and core values of the organization, which would delimit the boundaries of legitimate ICANN actions and decisions;

3. an Independent Review Mechanism, resembling juridical review, which would provide subjects affected by ICANN's decisions with appeal and redress procedures guided by ICANN's mission, commitments, and core values;

4. the ICANN board, representing, in this view, the executive entity held accountable to the community through the community's powers.

Following the state analogy, the articles and sections establishing the missions, commitments, core values, the empowered community, and the independent review were declared part of the "Fundamental Bylaw," which could be changed only with a supermajority vote of three-quarters from both the ICANN board and the ICANN community, like a rigid constitution. Nevertheless, in this proposal, ICANN remained a nonprofit organization (or public benefit corporation) under the jurisdiction of the California Corporations Code. This framework was also maintained in the following two draft proposals, published on August 3 and November 30 (followed by public comment periods form August 3 to September 12 and from November 30 to December 21, respectively).[23] The third proposal incorporated the previous discussion as twelve recommendations to be implemented into a reworked ICANN bylaw.

The comments generally welcomed the effort to enhance ICANN's accountability and agreed with the overall approach of the proposal, while most of the discussion concerned the details of how to implement these principles. Substantial issues were raised, including the jurisdiction over ICANN, the accountability of ICANN staff and supporting organizations/advisory committees, the accountability of ICANN toward subjects external to the ICANN community, the configuration of the empowered community under the California Corporate Code, and the role and

---

[23] See    https://www.icann.org/public-comments/gnso-review-2013-07-15-en    and https://www.icann.org/public-comments/draft-ccwg-accountability-proposal-2015-11-30-en. Accessed 12 June 2020.

powers of the GAC in the renewed governance structure (please see Chapter 7 for details).

To finalize the proposal, further reviews including both participants and the legal counsels were carried out during January and February 2016. The final proposal was distributed to the chartering organizations for approval on February 23, with minority statements submitted by 5 of the 28 voting members of the CCWG-Accountability, each belonging to a different chartering organization.

At the beginning of March, all of the chartering organizations approved the final report. Only the GAC response specified that no consensus had been reached on recommendation 11 and some parts of recommendations 1 and 2 concerning GAC prerogatives. However, no objection was raised against the transmission of the proposal to the ICANN board.

## 4.4  APPROVAL AND IMPLEMENTATION BY THE ICANN BOARD AND NTIA

On March 10, during the Marrakesh meeting, the accountability proposal was submitted to the ICANN board. On the same day, the board verified that both the "IANA Stewardship Transition Proposal" and the "Enhancing ICANN Accountability Recommendations" met the NTIA criteria, and decided to transmit the documents to the NTIA.[24] At the same time, the board launched an implementation plan aimed at ensuring that all the necessary changes were put in place by the expiration date of the IANA function contract (September 30, 2016), to be ready in the case of approval of the transition proposal by the NTIA. As most of this work required amending the ICANN bylaw, the Bylaw Coordination Group was established during the Marrakesh meeting, and in a joint effort with the ICANN team and the external counsels of the CCWG-Accountability and CWG-Stewardship, a new bylaw was drafted and published on April 21, available for public comment until May 21.[25] Unfortunately, information about the membership, charter, and engagement rules of the Bylaw Coordination Group have not been made publicly available.

---

[24] See    https://icann.org/resources/board-material/resolutions-2016-03-10-en#2.b. Accessed 12 June 2020.

[25] See    https://www.icann.org/en/system/files/files/report-comments-draft-new-bylaws-25may16-en.pdf. Accessed 12 June 2020.

Once it was ensured that the revised bylaw reflected the transition proposal's provisions and recommendations, the new bylaw was approved on May 27, 2016, specifying that it would come into force upon the expiration of the IANA function contract.[26] On June 9, 2016, the NTIA released its assessment report on the IANA transition proposal, testifying that it had met the criteria outlined in its March 2014 announcement and that it was consistent with good governance principles and best practices. After a further scrutiny by the US Congress, the NTIA let the IANA function contract expire on September 30, 2016, allowing the new post-transition arrangements to come into effect.

---

[26] See    https://www.icann.org/resources/board-material/resolutions-2016-05-27-en. Accessed 12 June 2020.

CHAPTER 5

# The Input Legitimacy of the IANA Transition Process

**Abstract** The chapter assesses the consistency of the IANA transition process with the principle of input legitimacy considered alongside the dimensions of "inclusiveness," "balanced representation," and "representativeness." Besides a traditional stakeholder analysis, an affiliation network analysis is performed in order to reach an in-depth and detailed portrait of the constellation of actors and interests that the participants brought into the decision-making process, as well as to investigate revolving doors, blurring boundaries and hidden power structures among stakeholders. Findings show that the selection and categorization of stakeholders reproduced within the IANA transition process the same misrepresentations and power imbalances already existing in the ICANN governance structure.

**Keywords** IANA transition · ICANN · Input legitimacy · Stakeholder analysis · Affiliation network analysis

© The Author(s) 2021                                                                 81
N. Palladino and M. Santaniello, *Legitimacy, Power, and Inequalities in the Multistakeholder Internet Governance,*
Information Technology and Global Governance,
https://doi.org/10.1007/978-3-030-56131-4_5

## 5.1    ANALYZING INPUT LEGITIMACY:
## OBJECTIVES, DATA, AND METHODS

This chapter assesses the consistency of the IANA transition process with the principle of input legitimacy. As described in Chapter 2, input legitimacy could be broken down into three dimensions: inclusiveness, balanced representation, and representativeness.

Therefore, this chapter aims at answering the following questions: Did the institutional design of the IANA transition process involve all relevant actors and interests in the decision-making process? Have been stakeholders represented in a balanced manner? Were participants in the decision-making process genuine representative of their respective stakeholder groups?

Responding to these questions requires taking into account how the stakeholders' categorization, selection, and identification procedures settled-up for the IANA transition process have affected the composition of the groups charged with the drafting of the transition proposal. We know from Chapter 2 that the categorization and selection of stakeholders, besides being *per sé* an inclusion/exclusion principle, also impacts the balancing of different actors and their possibility to communicate their views. Moreover, stakeholders' categorization and selection procedures are an even more fundamental source of institutional power since they tend to become taken for granted as soon as they are settled.

As seen in Chapter 4, with its announcement, NTIA unilaterally entrusted ICANN with the mandate to initiate a global multistakeholder process in order to come up with a transition proposal. In so doing, ICANN has been in the position to exert a high, and for a large degree arbitrary, power concerning the process design. In this regard, the first proposal released by the ICANN board in April 2014 envisaged the establishment of a working group composed by representatives of ICANN's SO/AC structures appointed by the Board itself with the GAC. Such arrangement resembled a "club governance" structure where who is already a member of the club "place a limit to the range of actors involved in the making of policy and define what type of actor is relevant" (Tsingou 2015: 231), and seemed to reveal the Board's intent to steer the process. Indeed, this choice gave rise to a strong reaction among commenters, which complained about the ICANN conflict of interest, and asked for a more inclusive and bottom-up process going beyond the ICANN ecosystem's borders. These criticisms have also arisen within the ICANN community and especially among the less represented components in the ICANN multistakeholder structure (civil society, government, business community not involved in registry operations), bringing to light an internal dividing line between, on the one hand, name, address, and protocol registries and, on the other hand, all the remaining stakeholder groups. In its revised proposal, the ICANN board made efforts to accommodate some of the claims raised during the public comment period. It turned back on its purpose to appoint the members of the coordination group and allowed each stakeholder group to select its representatives based on its own rules and procedures. Moreover, it added some further seats for actors external to the ICANN system and let the coordination group define its scope and working methods.

Nevertheless, the ICANN board revealed a high degree of arbitrariness in dealing with the collected feedback. While requests for a balance coming from the ICANN community or ICANN customers were, to some degree, satisfied, developing countries' claims for an equal representation obtained just the inclusion of "diversity" among the principles inspiring the transition, but no formal mechanism to ensure it. Also, views calling for alternative stakeholder categorizations and the establishment of an ad hoc multistakeholder initiative independent of ICANN were not taken into account.

The adopted changes did not alter the backbone of the process as imagined by the ICANN board for the IANA transition, which fundamentally remained based on the SOs/ACs ICANN structure. Overall, this kind of approach seems to indicate that ICANN identified its own community as the relevant global multistakeholder community.

Later, the IANA Transition Coordination Group (ICG) further concentrated decisional power in the hands of the subjects already responsible for the management of the IANA functions (namely ICANN and its contractual partners IETF and NRO), deciding to entrust these operational communities with the drafting of the transition proposal.

This chapter aims at investigating how such categorization and selection procedures affected the input legitimacy of the IANA transition process in its three dimensions of inclusiveness, balanced representation, and representativeness. In particular, this analysis will verify if the process design reproduced, within the IANA transition process, power relationships, well-known representation bias, and power imbalances already existing in DNS management.

To this aim, we traced back the profile of the 90 decision-makers with voting power included into the four ad hoc bodies established to draft the transition plan: the IANA Stewardship Transition Coordination Group (ICG), the Cross-Community Working Group on Naming Related Functions (CWG-Stewardship),[1] the Consolidated RIR IANA Stewardship Proposal Team (CRISP Team), the Cross-Community Working Group on Enhancing ICANN Accountability (CCWG-Accountability). We did not consider the IANAPLAN group settled by the IETF since, unlike the other cases, it was not based on official memberships with voting rights.

To this purpose, we collected information about member's nationality and career resorting to: (i) the biographic note and Statement of Interest (SOI) that the 90 members submitted within the IANA transition process, (ii) their ICANN wiki pages, and (iii) search engines.

The information thus obtained has been employed to analyze and compare the distribution of the 90 IANA transition voting members among different stakeholder categories.

Moreover, we performed an affiliation network analysis considering not only the official and declared affiliations of their decision-makers, but also all their other affiliations during the transition.

Affiliation network analysis is a particular kind of social network analysis, a family of techniques employed within the stakeholder analysis field in order to "organize data on the relational ties linking stakeholders together' and to 'uncover the structure of the stakeholder network" (Reed et al. 2009: 1939). Affiliation network analysis is designed to explore two-mode networks, namely networks composed of two different sets of entities (Borgatti et al. 2013; Carrington et al. 2005), in this case, members and organizations. While this technique has been scarcely considered in the study of multistakeholder governance and Internet governance, it has been widely and successfully employed in the study of corporate governance, especially for what concerns the phenomenon of interlocking directorates (Mizruchi 1996; Davis et al. 2003). This latter concept refers to overlapping memberships among boards of directors within a national or sectorial set of corporations that may favor the diffusion of information and knowledge (Chua and Petty 1999), promote

---

[1] In truth, CRISP Team had just ten voting members and five RIRs staff members (one for each RIR). However, considering the contributions of staff members within such a small body and close-knit community, we also included staff members in the set of the IANA-Transition decision-makers.

coordinated action among a group of actors, and allow "powerful and influential firms to exercise control over others" (Sankar et al. 2015; Seidel and Westphal 2004). Instead of analyzing co-membership among the boards of a pre-established set of organizations, our approach focuses on the individual members of the IANA transition decision-making bodies and re-traces their multiple affiliations. To this purpose, through the software Gephi, we extracted from a matrix members-organizations a network of organizations, where a tie between two different organizations corresponds to a shared member, and then we created related graphs and statistics. Affiliation network analysis allowed us to achieve a more in-depth and clearer portrayal of the constellation of organizations and interests that participants brought in the IANA transition process.

Moreover, affiliation network analysis enabled an investigation of how the revolving-door phenomenon may have blurred the boundaries among different stakeholder categories. This latter aspect has relevant implications both for inclusiveness, balanced representation, and representativeness. Multistakeholder initiatives rest upon the assumption that different stakeholders will bring a different perspective on the issue. Cross-sectorial memberships could mean that "given their previous similar socialization, the participants in these initiatives have similar—not different—ideas and resources, in the form of expertise, and that they share causal beliefs and practices," limiting the point of views included in the process, and placing dividing lines "within the different sectors rather than between them" (Schneiker and Joachim 2018: 10).

At the same time, multiple cross-sectorial affiliations may raise questions about which interests and constituencies have been actually represented by participants.

Further, affiliation network analysis allowed us to detect hidden power structures that usual analysis of formal membership leaves uncovered. Indeed, it makes observable: (1) which actors and stakeholders groups benefitted of multiple channels of access to the process; (2) the most influential actors, according to their connections and the potential ability to spread their views and ideas; and (3) pre-existing networks of participants and organizations that could be considered coalitions capable of steering the process.

## 5.2   Inclusiveness and Balanced Representation

Concerning the criteria of inclusiveness and balanced representation of stakeholders, a first analysis of the members' profiles seems to suggest that the IANA transition process has been surely inclusive but highly unbalanced, reproducing well-known asymmetries of the DNS regime within the drafting process.

Table 5.1 contains the distribution of the voting members among different stakeholder categories. The left side of the table shows the number of voting members among the IANA/ICANN official stakeholder groups. On the right side, we classified members resorting to the categories commonly used within multistakeholder governance studies and the WSIS/IGF ecosystem, including technical community, governments, civil society, and private sector, to which we added the category of "technical operators" (see below for details). To this purpose, we retraced the chain of affiliations leading each member to be appointed to one of the transition's working groups (e.g., a member was appointed to ICG since he/she belongs to ALAC as a member of a civil society association). On this basis, we assigned each member to a stakeholder category.

**Table 5.1**  Distribution of voting members among IANA/ICANN constituencies and sectorial categories

| IANA Constituencies | N | % | Sectorial categories | N | % |
|---|---|---|---|---|---|
| NRO | 17 | 18.9 | Technical Operators | 36 | 40.0 |
| ccNSO | 14 | 15.6 | Of which: | | |
| GNSO | 13 | 14.4 | Technical Operator (Private Sector) | 19 | |
| ALAC | 12 | 13.3 | Technical Operator (Mixed) | 15 | |
| GAC | 12 | 13.3 | Technical Operator (Governmental) | 2 | |
| SSAC | 6 | 6.7 | | | |
| ASO | 5 | 5.6 | Private Sector | 14 | 15.6 |
| gTLD Registries | 2 | 2.2 | Technical Community | 14 | 15.6 |
| IAB | 2 | 2.2 | Governments | 13 | 14.4 |
| IETF | 2 | 2.2 | Civil Society | 5 | 5.6 |
| ISOC | 2 | 2.2 | N.d. | 8 | 8.9 |
| RSSAC | 2 | 2.2 | | | |
| ICC/Basis | 1 | 1.1 | | | |
| Total | 90 | 100.0 | Total | 90 | 100.0 |

*Source* Authors' Creation

According to both the stakeholder classifications, the IANA transition process could be considered inclusive, meaning that the bodies entrusted to draft the transition proposal included all the considered stakeholder groups. Moreover, also following Mena and Palazzo's (2012) suggestion to take into account the number of criticisms regarding the exclusion of some stakeholder categories, the process still seems satisfactory.

As aforementioned, after the strong reaction to its first proposed process design, the ICANN board accommodated the requests of additional ICG seats for business community members and top-level domain operators (also ccTLD) not belonging to ICANN structures.

After these changes, the successive ICG decision to assign the drafting of the transition proposal to the three operational communities operating around ICANN, IRTF, and NRO encountered far less resistance. However, it is worth noting that CWG-Stewardship reserved voting power only to SOs/ACs members and that CRISP Team was composed only by RIRs' members.

Similarly, claims for extending participation beyond the ICANN community in the Enhance Accountability process were strongly mitigated as soon as the board accepted the establishment of a Cross-Community Working Group open to all interested parties, even if voting rights were reserved to SOs/ACs appointed members only.

Instead, the analysis of the balance among the different stakeholder groups leads to a negative assessment and points out relevant differences among the two categorizations.

Following the official IANA/ICANN classification, RIRs and address registries, through both NRO and ASO, are largely the most represented group with 22 members. Further, ccNSO, GNSO, ALAC, and GAC gained between 12 and 14 appointees each. The technical communities (SSAC, RSSAC, IETF, IAB, ISOC) got a similar amount of representatives. In contrast, representatives outside the ICANN ecosystem gained a minimal presence.

If we consider only the ICANN constituencies, this distribution returns a certain degree of equilibrium among them. Nevertheless, it is worth noting that most of the voting members represent address or name registries (42). If we also add the appointees of the other technical communities, it could be said that a vast majority (56) of the people who have taken part in the drafting of the IANA transition proposal are bearers of technical and operative interests. Against this overwhelming presence of the technical community, the representation of governments (twelve seats through the GAC); end-users (twelve seats by ALAC and three by the Non-Commercial Users Constituency of GNSO); and business constituencies (seven seats among ICC/Basis, GNSO's Commercial Stakeholder Group, and Registrars Stakeholder Group) appears much less meaningful.

An even more unbalanced representation comes to light if we distribute the 90 voting members of the IANA transition among stakeholder categories commonly employed within multistakeholder studies and the WSIS ecosystem.

As mentioned above, we have taken into account the affiliation external to ICANN that justified the inclusion of each member in some ICANN/IANA structure. On this basis, we have classified each member as private sector, governments, technical community, civil society, and technical operator. This latter label refers to organizations such as address and name registries, network information centers, Internet exchange points, which: (1) are constituted as no-for profit organizations, usually grouping different kinds of actors and headed through a multistakeholder governance structure; (2) carry out technical operative functions related to the DNS and provide services to their affiliate under the payment of a fee.

Academic research centers, not directly under governmental control, also fall into this category. In other words, this category differs, on the one hand, from technical communities which develop protocols, standards, guidelines, and the like, but do not offer services, and on the other hand, from service providers constituted as a business company or a governmental authority (which have been classified respectively as private sector and government). Further, we scrutinized statutes, by-laws, and membership of the technical operators' organizations in order to understand which kind of interests they represent.

Following this alternative categorization of stakeholders appears that the drafting process of the IANA transition had been dominated by an interweaving of technical and economic interests, with a significative presence of governments, while civil society constituted a very marginal voice.

Indeed, 36 of the 90 IANA transitions' voting members belong to technical operator organizations. Despite their not-for-profit nature, in most cases (19 out of 36), membership and steering bodies of such technical operators are composed almost entirely by business actors. Besides this, private sector organizations and technical communities had 14 voting members each; governments 13, and civil society only 5. Eight members could not be classified with these categories. For the most part, they are members appointed by SSAC on an individual basis, with no relation with any further constituency or organization.

A point deserving closer attention is the underrepresentation of civil society, compared with what should be expected based on the IANA/ICANN categorization. Indeed, ALAC is supposed to represent individual end-users and appointed twelve members. Moreover, also GNSO includes non-commercial users' constituency. Indeed, the number of members who were an expression of civil society associations appears surprisingly low. This underrepresentation of the civil society is mainly because ALAC sub-structures often appointed people who were directors or presidents of regional/national ISOC chapters, usually with relevant working experience within registries or ICT companies.

Even the analysis of the distribution of the 90 IANA transition members among geographical and geopolitical categories reveals strong asymmetries (Table 5.2).

Regarding nationality, Western countries appear to be over-represented within the drafting and decisional organism involved in the IANA transition process. In particular, US citizens constitute the most remarkable group, occupying 20 seats over 90 available.

The aggregation of data based on broader geographic areas confirms this first observation. Indeed, Europe and North America appear to be

**Table 5.2** Distribution of the 90 IANA transition members among geographical and geopolitical categories

| Nationality | N | % | Geographic area | N | % | Development | N | % |
|---|---|---|---|---|---|---|---|---|
| USA | 20 | 22.2 | Europe | 26 | 28.9 | Advanced Economies | 59 | 65.6 |
| Australia | 7 | 7.8 | North America | 24 | 26.7 | Developing Countries | 26 | 28.9 |
| Netherlands | 5 | 5.6 | Africa | 13 | 14.4 | BRIC | 5 | 5.6 |
| Argentina | 4 | 4.4 | South America | 12 | 13.3 | | | |
| Canada | 4 | 4.4 | Oceania | 9 | 10.0 | | | |
| Germany | 4 | 4,.4 | Asia | 6 | 6.7 | | | |
| ... | ... | ... | | | | | | |
| Total | 90 | 100.0 | | 90 | 100.0 | | 90 | 100.0 |

*Source* Authors' Creation

the most represented regions, with 24 and 26 seats each, amounting to the 55.6% of the total. Africa, South America, and Oceania appointed 13, 12, and 9 members, respectively. Most notably, Asia, a continent with about two billion Internet users, is the least represented region with just six voting members appointed in all the drafting groups (North American and Europe together have less than a billion Internet users).

A further aggregation based on the developing stage portrays an even more unbalanced picture. We divided countries in Advanced Economy (including USA, Canada, Europe, Japan, Australia), BRIC (Brazil, Russia, India, China), and Developing Countries (including the countries not within the previous categories). Advanced Economies gained over 65% of the IANA transition voting members, while BRIC and Developing Countries got, respectively, only 5% and 26%.

Overall speaking, the results of this stakeholder analysis seem to indicate that the adopted categorization and appointment procedures have reproduced within the IANA transition process well-known power relationships and imbalances already existing in the DNS management, over-representing Western, technical, and business interests while marginalizing developing countries and civil society participation. It is worth noting that governments appear to gain a significant representation, continuing that trend of gradual empowerment we have discussed in Chapter 3. In the first version of the process design, GAC was the only group allowed to select its own representatives, and during the process, ICG increased GAC's members from 2 to 5, testifying of ICANN efforts to ensure an international recognition to the transition process.

## 5.3   Representativeness, Revolving Doors, and Blurring Boundaries Among Stakeholders' Groups

As seen in Chapter 2, another relevant criterion to assess input legitimacy is representativeness. In the first place, representativeness refers to internal accountability governing the relationship among a participant of a decision-making body and the constituency that appointed him (Risse 2006). It requires considering the mechanisms through which a stakeholder group could ensure that their representatives act according to its will and interest.

In our view, this means take into account three different kinds of mechanisms: mechanisms through which constituency choose their representatives; mechanisms through which the representatives report their activities to their group; and mechanisms through which a constituency could remove its representatives in the case of disappointing or disloyal behavior.

As mentioned above, after an initial attempt by the ICANN board to arrogate to itself the selection of the members of the ICG, the various constituencies involved obtained to choose their own representatives by themselves, and they maintained such prerogative also in the other drafting bodies established lately.

Instead, the charters of ICG, CWG-Stewardship, CCWG-Accountability, and CRISP Team do not mention reporting duties of their members toward the respective communities, neither removal procedures. Rather, they all just encourage members to exchange views with their respective appointing organizations.

However, there is no trace of conflicts between members of the drafting bodies and their respective communities in publicly available documentation about the transition. Moreover, ICG, CWG-Stewardship, and CCWG-Accountability charters provided that the chartering organizations should have approved the output of the respective bodies. Fail in reaching a consensus would have been documented in the final report. In particular, the ICG charter established that the objection of a majority of an operational community would have precluded the ability of the ICG to submit an acceptable consensus proposal to the Board. However, within every single operational community, the opposition of a single constituency, or a small minority of constituencies, would not have necessarily precluded the finalization of the process.

In sum, the constituencies/chartering organizations had greater control over the output of the process than on their representatives.

Such uncertainty about the internal accountability mechanisms leads to paying attention to the other and broader meaning of representativeness, which takes into account to what extent participants involved in a decision-making process, as well as their constituencies, actually represent the kind of actors in the behalf of whom they are supposed to act. According to this second conception, the assessment of the representativeness of a process requires going beyond the formal procedures governing the relationship between representatives and constituencies. Instead, more substantially, the analysis should pay attention to the network of affiliations of each participant and the constituencies' capability to voice the category they claim to represent.

In this view, it is essential to detect if revolving doors phenomena have taken place and undermined the assumptions at the basis of the input legitimacy of the multistakeholder model, i.e., the possibility to assemble actors from different backgrounds bearing different points of view and expertise. The more the boundaries among different stakeholder groups are blurred, the more the supposed diversity and complementarity among stakeholders disappears, and the multistakeholder decision-making bodies end up resembling like-minded global élite.

Then, if most of the representatives of a stakeholder group are also members of other stakeholder groups, this could lead to talk about misrepresentation. Further, if the vast majority of the decision-making members of a multistakeholder initiative belong to one or a few categories, it could be said that such categories have dominated the process.

In order to explore this dimension, we performed an affiliations network analysis taking into account all the affiliations of the 90 decision-makers during the IANA transition process. Given the higher number and variety of the organizations included, we added some further categories, already recognized in Internet governance multistakeholder studies and practices, namely academic community, international organizations, and multistakeholder dialogue fora.

In this section, we will consider only data related to the composition of this extended network and the overlapping among the different stakeholder categories. In the next section, we will expose network statistics to explore the power relationships that structure the network.

In the first instance, Table 5.3 confirms the imbalances in the IANA transitions decision-making bodies previously observed: 50 of the 90 voting members of the IANA transition process were affiliated to at least one of the 31 organizations classified as technical operators; the private sector groups 38 voting members and the highest number of organizations (49); the technical community had 39 voting members in just four organizations (33 of them in the only ISOC); governments collected 22 members in 21 different organizations, while other stakeholder categories had a less consistent presence.

Table 5.4 and Fig. 5.1 point out that the multiple affiliations of the IANA transition decision-makers led to a substantial overlap among different categories of actors. Most of the 90 members belong to technical operators, technical communities, or the private sector. Table 5.4

**Table 5.3** Composition of extended network

| | N. voting member affiliated | N. organizations | | N. voting member affiliated | N. organizations |
|---|---|---|---|---|---|
| Official ICANN/IANA constituencies | 69 | 16 | Multistakeholder forum | 16 | 7 |
| Technical operator | 50 | 31 | Academic | 15 | 23 |
| Technical community | 39 | 4 | Civil society | 10 | 14 |
| Private sector | 38 | 49 | Intergovernmental organization | 9 | 8 |
| Government | 21 | 22 | Other | 7 | 6 |

*Source* Authors' Creation

**Table 5.4** Overlapping rate between IANA constituencies and stakeholder types

| | Academic | Civil society | Govern. | Internat. org. | Multistak. fora | Private sector | Tech. com. | Tech. op. | Other |
|---|---|---|---|---|---|---|---|---|---|
| ALAC | 0.29 | 0.21 | 0.07 | 0.14 | 0.21 | 0.57 | 0.79 | 0.50 | 0.50 |
| GNSO | 0.20 | 0.10 | 0.00 | 0.00 | 0.00 | 0.85 | 0.25 | 0.35 | 0.15 |
| GAC | 0.07 | 0.00 | 0.93 | 0.29 | 0.29 | 0.14 | 0.21 | 0.21 | 0.07 |
| ccNSO | 0.13 | 0.07 | 0.20 | 0.07 | 0.13 | 0.27 | 0.13 | 0.93 | 0.00 |
| ASO | 0.00 | 0.00 | 0.14 | 0.14 | 0.29 | 0.14 | 0.57 | 1.00 | 0.00 |
| SSAC | 0.43 | 0.29 | 0.14 | 0.00 | 0.00 | 0.29 | 0.57 | 0.43 | 0.00 |
| RSSAC | 0.50 | 0.00 | 0.00 | 0.00 | 0.00 | 0.00 | 1.00 | 1.00 | 0.00 |
| NRO | 0.20 | 0.05 | 0.20 | 0.05 | 0.25 | 0.30 | 0.50 | 1.00 | 0.05 |
| IETF | 0.00 | 0.00 | 0.00 | 0.00 | 0.00 | 0.80 | 1.00 | 1.00 | 0.00 |
| IAB | 0.00 | 0.33 | 0.00 | 0.00 | 0.33 | 1.00 | 1.00 | 0.00 | 0.00 |
| ISOC | 0.30 | 0.24 | 0.18 | 0.15 | 0.30 | 0.42 | 1.00 | 0.64 | 0.09 |
| Academic | 1.00 | 0.33 | 0.13 | 0.07 | 0.20 | 0.33 | 0.67 | 0.60 | 0.07 |
| Civil society | 0.50 | 1.00 | 0.10 | 0.10 | 0.30 | 0.50 | 0.80 | 0.40 | 0.10 |
| Governments | 0.10 | 0.05 | 1.00 | 0.24 | 0.19 | 0.10 | 0.29 | 0.43 | 0.10 |
| Internat. org. | 0.11 | 0.11 | 0.56 | 1.00 | 0.33 | 0.33 | 0.56 | 0.33 | 0.11 |
| Multist. fora | 0.19 | 0.19 | 0.25 | 0.19 | 1.00 | 0.19 | 0.63 | 0.56 | 0.06 |
| Private sector | 0.13 | 0.13 | 0.05 | 0.08 | 0.08 | 1.00 | 0.47 | 0.34 | 0.11 |
| Tech. com. | 0.26 | 0.21 | 0.15 | 0.13 | 0.26 | 0.46 | 1.00 | 0.56 | 0.18 |
| Tech. op. | 0.18 | 0.08 | 0.18 | 0.06 | 0.18 | 0.26 | 0.44 | 1.00 | 0.08 |
| Other | 0.14 | 0.14 | 0.29 | 0.14 | 0.14 | 0.57 | 0.43 | 0.57 | 1.00 |

*Source* Authors' Creation

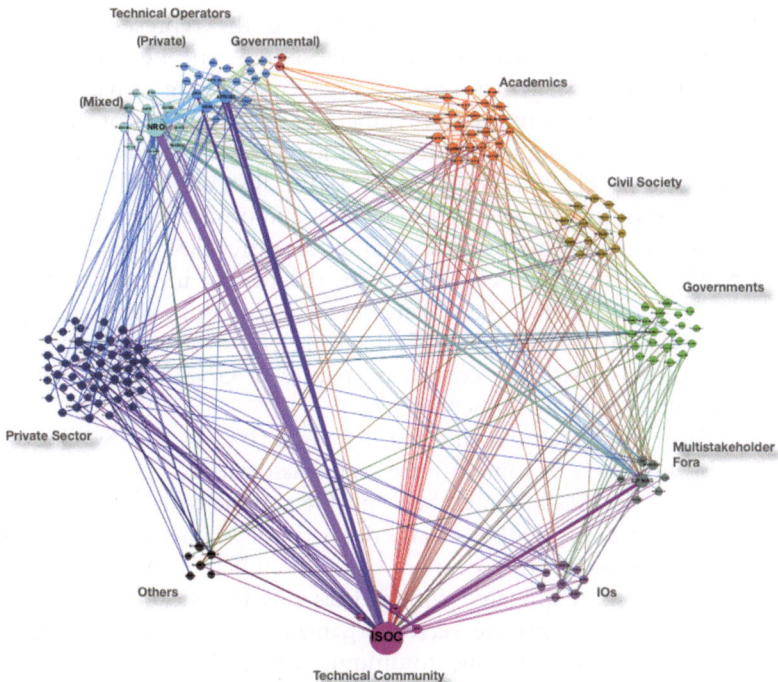

**Fig. 5.1**  Overlapping membership among IANA transition stakeholders (*Source* Authors' Creation)

also shows that most of the academic community, civil society, governments, and international organizations members were also affiliated with technical operators, technical communities, or private sector, while the opposite was not true. For example, 50% of members affiliated to a civil society association also belong to a private sector association, 80% to a technical community association, and 40% to a technical operators' organizations. By contrast, members of civil society constitute 13%, 21%, and 8% of the private sector, technical community, and technical operator organizations, respectively.

Further, even technical operators, the technical community, and the private sector are significantly interwoven. As we will explore much in detail in the next section, they constitute the power structure of the network.

These overlapping affiliations undermined the representativeness of some constituencies since members appointed on the basis of their membership of a particular type of organization (e.g., civil society or advocacy) are at the same time members of organizations of other kinds with different if not opposite interests. Indeed, while the profiles of some constituencies as GAC, ccNSO, ASO, SSAC, RSSAC, NRO appear well-defined and consistent with their expected composition, the same cannot be said of other stakeholder groups and constituencies.

GNSO shows a well-defined profile; the people in this constituency belong for the 85% to some private sector association, while other categories have a much lower level of affiliation. Nevertheless, this data seem hard to reconcile with the composite nature of GNSO that is supposed to represent also non-commercial actors.

People in the IANA transition affiliated to IETF/IAB/ISOC are very often also members of some private sector association.

Above all, the ALAC's members' profile appears to be the most problematic, whereas it is neither clearly defined nor in line with expectations based on its mission. Indeed, most of the members of ALAC within the IANA transition are affiliated with at least one technical operators, technical community, or private sector organization. By contrast, only 29% also belong to some academic community and 21% to some civil society association. This data raise some doubts about ALAC representativeness, considering that this constituency is supposed to represent Internet end-users and the global public interest.

## 5.4   POWER RELATIONSHIP IN THE "SMALL WORLD" OF THE IANA TRANSITION NETWORK

Table 5.5 summarizes some general proprieties of the affiliation network (Fig. 5.2), namely: (1) the number of nodes, that is the number of organizations; (2) the number of edges, that is the number of links among organizations; (3) the average degree, that is the average number of nodes each node is connected with; (4) the network diameter, that is the maximal possible distance between two nodes within the same network measured in edge-hops; (5) graph density, that is equal to the number of links in the network divided by the number of potential links; (6) the clustering coefficient, that is a measure of the tendency of the nodes to group into clusters; and (7) the average path length, that is the average distance among pairs of nodes within the network measured in edge-hops.

**Table 5.5** Affiliation network properties

| Nodes | 180 |
|---|---|
| Edges | 624 |
| Average Degree | 6.933 |
| Network diameter | 5 |
| Graph density | 0.039 |
| Average clustering coefficient | 0.872 |
| Average path length | 2.546 |

*Source* Authors' elaboration using Gephi software

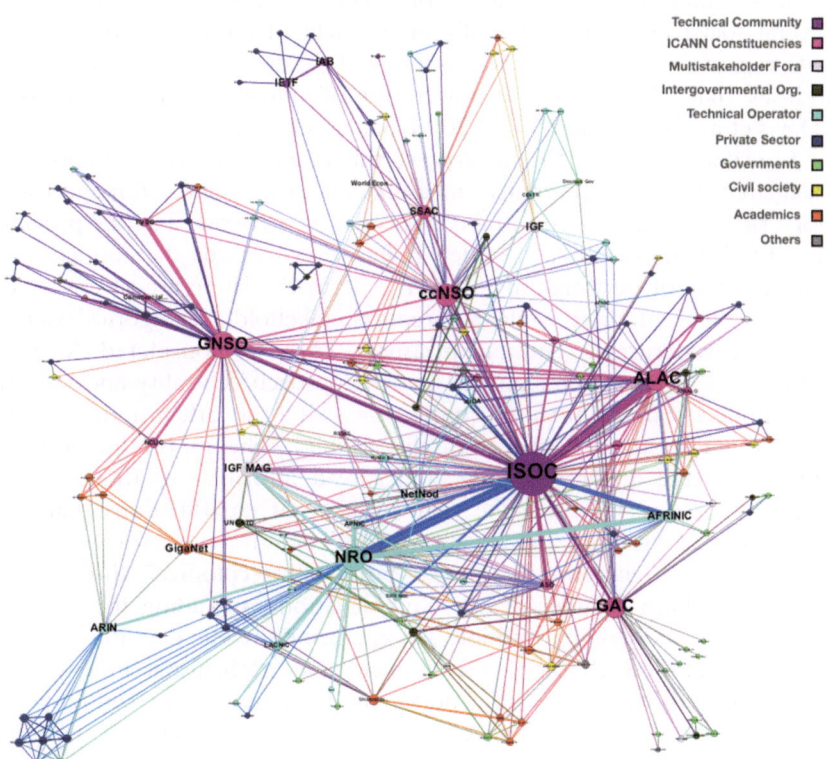

**Fig. 5.2** The IANA transition affiliation network (*Source* Authors' Creation)

Network statistics show the typical features of the so-called small-world network (Watts and Strogatz 1998; Robins and Alexander 2004), namely low graph density, small network diameter, short average path lengths, and a high average clustering coefficient. Small-world networks are networks where the mean distance between pairs of nodes grows less than proportionally to the logarithm of the number of nodes N in the network. In other words, small-world networks are networks where almost every node is connected to another through a relatively short chain of intermediaries. Small-world structures have been detected in several social networks, especially in the field of corporate governance, with regard to the phenomenon of interlocking directorates. Usually, they indicate that a homogeneous power élite of influential actors is at work (Davis et al. 2003; Baum et al. 2003; Sankar et al. 2015).

In order to identify the most powerful actors within the network, we calculated different "centrality" measures for each organization and category of organizations. Centrality measures in network analysis provide quantitative scores on each node's relevance as regards the connectivity and the flow of information within the network. There are several ways to calculate centrality; Table 5.6 reports stakeholder categories' values for three widespread centrality measures. First, we considered "degree centrality," which is the simplest way to approach centrality and consist of a node's total number of connections with the other nodes in the network. The more degree centrality is high, the more the node is well connected and can acquire and spread information. In our case, degree centrality indicates the total number of co-affiliations that the members of each node/organization have. Then, if a node/organization has a high degree centrality, it may potentially "colonize" through its members a high number of other organizations, spreading its views in the network and gaining multiple access channels to the decision-making process, regardless of official constituencies and mechanisms. Eigenvector centrality weights the ties of a node based on the degree centrality of the node to which it is connected and could be considered a measure of how meaningful the connections of a node are. Lastly, betweenness centrality measures "how often a given node falls along the shortest path between two other nodes" and it is usually "interpreted in terms of the potential for controlling flows through the network" (Borgatti et al. 2013: 185). In our case, betweenness centrality indicates to what extent a node/organization acts as a bridge among different nodes and then could play a gatekeeper or coordination role in the dissemination of information and views across the network.

Data on centrality show that technical operators (especially NRO, ARIN, RIPE NCC, AFRINIC, APNIC, LACNIC, NetNod, CENTR) and technical community (above all ISOC, IETF, IAB) have been the most influential groups of actors in the network according to every measure. They have likely been able to spread their views across the network of participants and decision-making bodies. The technical community's role is even more notable considering that we have not taken into account in our analysis the IANAPLAN process, which is the part of the transition proposal explicitly assigned to the technical community.

In particular, ISOC is the single organization with the highest scores on degree, eigenvector, and betweenness centrality. It could be considered the "hub" of the process connected with almost half of all the organizations of the network.

Private sector organizations present a different pattern. As a group, they have a very high value concerning degree centrality. Instead, taken individually, they show a much lower score, meaning that members of private sector organizations usually have just a few other affiliations. Nevertheless, these connections appear to have relevant strategic value. As mentioned above, talking about Table 5.4, a vast majority of the members

**Table 5.6**  Centrality measures for stakeholder categories

| | Degree (sum) | Degree (mean) | Eigenvector centrality (mean) | Betweenness centrality (sum) | Betweenness centrality (mean) |
|---|---|---|---|---|---|
| Technical operator | 227 | 7.32 | 0.124302 | 3538.31 | 114.14 |
| Technical community | 106 | 26.50 | 0.292807 | 7509.91 | 1877.48 |
| Private sector | 197 | 4.02 | 0.070412 | 45.26 | 0.92 |
| Government | 78 | 3.55 | 0.068103 | 28.57 | 1.30 |
| Multistakeholder forum | 63 | 9.00 | 0.175281 | 1045.17 | 149.31 |
| Academic | 145 | 6.30 | 0.131796 | 481.73 | 20.94 |
| Civil society | 72 | 5.14 | 0.116005 | 0.00 | 0.00 |
| Intergovernmental organizations | 48 | 6.00 | 0.117092 | 106.91 | 13.36 |
| Other | 32 | 5.33 | 0.100306 | 0.00 | 0.00 |

*Source* Authors' elaboration using Gephi software

of GNSO, ALAC, IETF, and IAB are affiliated to some organization of the private sector. Then, private sector interests and concerns could have potentially found room in all these stakeholder groups.

Governments do not show a high value of centrality. They appear as a category concentrated in the GAC structure with a limited capacity to spread its views.

Civil society organizations have a low score on centrality measures, except for the eigenvector value, meaning that such organizations are at the periphery of the network; they are not crossroads connecting members of different organizations. Instead, their members tend to belong to a few more influential organizations with high degree centrality.

Finally, the academic community has relevant scores on all the centrality measures, while international organizations and multistakeholder fora are characterized by a relevant degree of betweenness, consistently with their composite and transnational venue nature.

Overall speaking, data from the affiliation network analysis render the picture of a web of intertwined and overlapping multiple memberships, which blurred the boundaries among stakeholders' categories, questioning the actual representation of interests by the different IANA/ICANN stakeholder groups. More in detail, data reveal the existence of a power structure in the affiliation network of the IANA transition process composed by a backbone of technical operator, technical community, and private sector. These organizations appear potentially able to propagate their views across the networks and to "colonize" other groups and most peripherical nodes, as well as to disseminate a common, pre-existent, understanding of how the transition should have been conducted.

## REFERENCES

Baum, A., Shipilov, A. V., & Rowley, T. (2003). Where Do Small Worlds Come From? *Industrial and Corporate Change, 12,* 697–725.

Borgatti, S., Everett, M., & Johnson, J. (2013). *Analyzing Social Networks.* London: Sage.

Carrington, P. J., Scott, J., & Wasserman, S. (2005). *Models and Methods in Social Network Analysis.* Cambridge: Cambridge University Press.

Chua, W. F., & Petty, R. (1999). Mimicry, Director Interlocks, and the Interorganizational Diffusion of a Quality Strategy: A Note. *Journal of Management Accounting Research, 11,* 93–104.

Davis, G. F., Yoo, M., & Baker, W. E. (2003). The Small World of the American Corporate Elite, 1982–2001. *Strategic Organisation, 1*(3), 301–326.

Mena, S., & Palazzo, G. (2012). Input and Output Legitimacy of Multistakeholder Initiatives. *Business Ethics Quarterly, 22*, 527–556.

Mizruchi, M. S. (1996). What Do Interlocks Do? An Analysis, Critique, and Assessment of Research on Interlocking Directorates. *Annual Review of Sociology, 22*, 271–298.

Reed, M. S., Graves, A., & Dandy, N. (2009). Who's In and Why? A Typology of Stakeholder Analysis Methods for Natural Resource Management. *Journal of Environmental Management, 90*, 1933–1949.

Risse, T. (2006). Transnational Governance and Legitimacy. In A. Benz & Y. Papadopoulos (Eds.), *Governance and Democracy* (pp. 179–199). London: Routledge.

Robins, G., & Alexander, M. (2004). Small Worlds Among Interlocking Directors: Network Structure and Distance in Bipartite Graphs. *Computational & Mathematical Organization Theory, 10*, 69–94.

Sankar, P. C., Asokan, K., & Kumar, K. S. (2015). Exploratory Social Network Analysis of Affiliation Networks of Indian Listed Companies. *Social Networks, 43*, 113–120.

Schneiker, A., & Joachim, J. (2018). Revisiting Global Governance in Multistakeholder Initiatives: Club Governance Based on Ideational Prealignments. *Global Society, 32*(1), 2–22.

Seidel, M. D. L., & Westphal, J. D. (2004). Research Impact: How Seemingly Innocuous Social Cues in a CEO Survey Can Lead to Change in Board of Director Network Ties. *Strategig Organisation, 2*, 227–270.

Tsingou, E. (2015). Club Governance and the Making of Global Financial Rules. *Review of International Political Economy, 22*(2), 225–256.

Watts, D. J., & Strogatz, S. H. (1998). Collective Dynamics of 'Small-World' Networks'. *Nature, 393*, 440–442.

# CHAPTER 6

# The Throughput Legitimacy of the IANA Transition Process

**Abstract** This chapter focuses on the throughput legitimacy of the IANA transition process. Throughput legitimacy refers to the "black box" of a governance system, particularly the legitimacy of the processes through which the different views, interests, and positions that participants bring to a multistakeholder initiative (or, more generally, in a decision-making process) are transformed into an outcome. The analysis takes into account the procedural quality of the IANA transition to assess if its institutional design gave equal and meaningful opportunities to all involved actors to participate and influence the outcome. Further, the discursive quality of the process is investigated, considering the extent to which the IANA transition was close to the ideal-type of deliberative procedure, and whether the process was flawed by hegemonic discursive practices that inhibited minority points of view.

**Keywords** ICANN · IANA transition · Throughput legitimacy · Fairness · Accountability · Discourse analysis

N. Palladino and M. Santaniello, *Legitimacy, Power, and Inequalities in the Multistakeholder Internet Governance*,
Information Technology and Global Governance,
https://doi.org/10.1007/978-3-030-56131-4_6

103

## 6.1    The Procedural Quality
## of the IANA Transition

Fairness is probably the critical component in the procedural quality of a decision-making process. In the first place, fairness means that "all stakeholders should be able to participate in the process either on an equal basis or on the basis of morally justified graduated participation rights" (Beisheim and Dingwerth 2008: 13). It can be divided into formal and substantial fairness (Hooker 2005), where the former requires the equal and impartial application of rules and procedures (Scholte and Tallberg 2018; Schmidt and Wood 2019), and the latter considers the effects produced by the institutional design of the process, the extent to which it ensures the "equality of opportunity to participate in a decision-making process in an adequate way" (Dingwerth 2007: 28), or provides an "equal say in decision-making" (Dellmuth et al. 2019: 13). In this view, "fair procedures are seen as unbiased and neutral, free from self-interested or ideological considerations" (Doherty and Wolak 2012: 303).

As we have seen in Chapter 4, a notable amount of time was employed in the negotiation of rules and procedures for the IANA transition process, often resorting to public consultation, sometimes leading to contestation and then to some sort of agreement. This kind of participatory activity in rules-writing is supposed to guarantee fairness, as it allows the rules to be discussed and scrutinized by participants. Despite these efforts, however, the process was characterized by shortcomings and controversial decisions, particularly at the beginning, producing a cascade effect.

The IANA transition process began with the NTIA entrusting ICANN to convene a global multistakeholder process and, at the same time, fixing several requirements for the final proposal to be approved. The US government took these decisions unilaterally without prior consultation with the global Internet community, despite the crisis of legitimacy it was experiencing after Snowden's disclosures.

The relinquishing of oversight over the IANA functions was also influenced by the heated debate the US Congress had on the transition. Republicans accused the Obama administration of "giving away" the Internet, and the approval of the DOTCOM bill subjected the adoption of the transition proposal to congressional scrutiny. In this political context, the definition of the transition through an ad hoc multistakeholder initiative (like Net Mundial), or even within a UN arena like IGF, would have probably produced strong political reactions in the USA, thus putting at risk the whole operation.

However, NTIA decisions introduced a twofold bias within the transition process, undermining its fairness. Not only had a single actor, ICANN, been entrusted with a leading position in shaping the process, but this actor was in a clear conflict of interest, as underlined by many comments in the preparatory phase. Contestation grew even stronger when the ICANN board released its framework proposal for the transition, putting ICANN's role as the IANA transition operator out of the scope of the process, and reserving the appointment of the members of the steering group to itself and the GAC. According to the commenters, these provisions constituted an unacceptable limitation of the discussion and representativeness of the process.

Despite these reactions leading the ICANN board to modify its original plan, the board still had the full power to deal with the comments and inputs collected during the public consultation. Although there were a few seats assigned to non-ICANN constituencies, the board succeeded in grounding the IANA Stewardship Transition Coordination Group mainly on ICANN's supporting organizations and advisory committees.

The decision of the ICG to entrust the "operational communities" around ICANN, the IETF, and the RIRS with the drafting of the transition proposal reduced the possibilities of participation for non-ICANN/IETF/RIRS members. Indeed, this framework created three different classes of stakeholders, with different degrees of participation:

1. stakeholders not belonging to the operational communities, who could only contribute to the process by providing inputs and comments;
2. stakeholders not affiliated to ICANN/IETF/RIRS but included in the ICG; and
3. members of the operational communities, in charge of approving their parts of the transition proposal and enjoying the broadest rights and opportunities in the decision-making process.

Moreover, it should be noted that the three operational communities arranged their decision-making processes in different ways. The parameters community (IETF), with the IANAPLAN working group, followed the usual "rough consensus and running code" approach with a process potentially open to all. The numbers community established a CRISP Team including just RIR members, while other types of actors could only observe the discussions of the CRISP Team via teleconference and read the related mailing list. It is worth noting that since RIRs are, for the vast majority, Internet service providers or Internet registries, this choice favored technical and commercial interests and undermined the role of actors bearing public, social, and political concerns related to the number function of the IANA. The names community distinguished attendees in the CWG-Stewardship between full "members," appointed by ICANN's supporting organizations and advisory committees, and "participants" including people beyond the chartering organizations, who could attend the CWG meetings and actively take part in the discussion, but not at "consensus call" and "decisions." The same rules and procedures were adopted by the CCWG-Accountability, even though ICANN considered the "enhanced accountability process" an internal ICANN process.

Were these differentiations among stakeholders justified? While the charters of the CRISP Team, the CWG-Stewardship, and the CCWG-Accountability did not provide any explanation for the exclusion of some stakeholders or differentiated participatory rights, the ICG charter provided a rationale for the primary role assigned to the operational communities.

In this regard, the discussion held at the ICG meeting on July 17–18, 2014 is enlightening.[1] It started by commenting on a draft of the charter circulated among ICG members, mainly based on the IAB submission to the April call for input. The IAB document introduced the theme of the division of the transition proposal into three components related to name, number, and protocol parameter functions, identifying ICANN, the RIRS, and the IETF as the respective communities of interest.[2] The special role accorded to the "operational communities" was justified by a pragmatic argument, according to which the procedure would have simplified and accelerated the process by relying on existing structures and decision-making procedures. Further, other participants maintained

---

[1] https://www.icann.org/en/system/files/files/transcript-chat-coordination-group-17j ul14-en.pdf; https://www.icann.org/en/system/files/files/transcript-chat-coordination-group-18jul14-en.pdf. Accessed 12 June 2020.

[2] https://www.iab.org/wp-content/IAB-uploads/2014/04/iab-response-to-201 40408-20140428a.pdf. Accessed 12 June 2020.

that the operational communities' interests are more relevant than others since they are involved in DNS functioning.

Only a few speakers suggested avoiding having references to the three operational communities. A more significant number of attendees pointed out that the IANA transition would affect a broader range of communities and stakeholders than operational communities and stressed the importance of moving beyond the ICANN system. The solution found by the ICG was to invite stakeholders external to the operational communities to submit their input within the related operational community processes. However, this choice created inconsistencies and discriminations among actors that are not morally justified nor logical. Stakeholders not belonging to the organizations at the core of the operational communities were called to join the process according to rules and procedures that they had not contributed to creating, and with minor participatory rights.

Another point of concern relates to the complexity of decision-making mechanisms. The official IANA transition site proudly claims that the process involved more than 800 working hours (while the NTIA calculated that stakeholders spent more than 26,000 hours!).[3] The IANA transition was discussed, organized, and planned in 33,100 mailing list exchanges, 600 calls and meetings of the major groups, and at more than 1100 events around the world.[4] These numbers document the amount of work, participation, and stakeholder involvement within the IANA transition process, as well as the seriousness and depth of this multistakeholder effort. Nevertheless, the overwhelming complexity of the process undermined the opportunities for participation for the weaker and less well-resourced stakeholders.

It should be evident that the two-year-long process, which required meetings every one or two weeks for each of the groups and subgroups, advantaged those organizations able to employ dedicated personnel to follow the process in its entirety (Hill 2016). Indeed, complexity creates discrimination among actors concerning their possibility to influence the outcome. As John C Klensin, the former IAB chair and emeritus member of the Internet community since its early stage, declared:

---

[3] https://www.ntia.doc.gov/blog/2016/reviewing-iana-transition-proposal.    Accessed 12 June 2020.

[4] https://www.icann.org/resources/pages/iana-accountability-participation-statistics-2015-11-04-en. Accessed 12 June 2020.

Very few people can afford to 'participate' at the level needed unless they have corporate or organizational support with very significant travel and salary (or equivalent) resources. Unsurprisingly, that level of support tends to come from organizations who have a vested interest [...] I note that the very ability to participate actively in 'over 100 calls and meetings and over 4,000 mailing list messages' excludes participants and, in all likelihood, stakeholder groups, by attrition and exhaustion.[5]

Data provided by the Center on Internet and Society on mailing activities confirm this view, suggesting that only 98 people contributed with more than 20 mails in the different official lists of the transition process, 12 of whom were members of ICANN staff while "76 were identifiable as primarily being from industry or the technical community."[6]

Besides causing time, financial, and organizational constraints to participation, complexity also acts on a cognitive level. The fragmentation of the outcome as several partial drafts, strictly interconnected and coming from different working groups and subgroups, made the proposal "hard to understand even by members and participants,"[7] to quote the minority statement of one CCWG-Accountability member. Again, only well-structured organizations, employing a proper staff, could manage the development of the process as a whole and act on every part according to their preferences.

Another relevant facet of procedural quality is accountability and the related criteria of transparency. In the previous chapter, we dealt with the "internal accountability" of the process, which refers to the mechanisms put in place to ensure that decision-makers remain accountable toward their constituencies. This chapter considers "external accountability," which "refers to people or groups outside the acting entity who are nevertheless affected by it" (Risse 2006: 7). According to Mueller (2009: 94), external accountability consists of "an oversight or appeals process conducted by an independent entity with the authority to reverse the organization's decisions or impose sanctions on it for misbehavior."

According to this view, there is no doubt that the external accountability of the IANA transition process was addressed primarily toward

---

[5] https://comments.ianacg.org/pdf/submission/submission119.pdf. Accessed 12 June 2020.

[6] https://comments.ianacg.org/pdf/submission/submission126.pdf. Accessed 12 June 2020.

[7] https://community.icann.org/pages/viewpage.action?pageId=53783460&preview=/53783460/54887671/Appendix%20H%20Minority%20Statements_FINAL.docx. Accessed 12 June 2020.

the US government, as its outcome was scrutinized first by the NTIA and then by the US Congress for the effect of the DOTCOM bill. This circumstance raises concerns about the legitimacy and accountability of the IANA transition as a global governance process, as the bodies entrusted to draft the transition proposal were accountable to the US government and Congress, besides (and maybe before) the Internet global multistakeholder community. As we will see later, this mechanism significantly influenced the discussion among members and participants, putting constraints on acceptable arguments and discourses.

Another form of external accountability is strictly related to transparency and refers to the relationship between the agent and the broader public of affected parties who, while devoid of any binding power, can pressure by contestation and "name and blame" tactics. Following Buchanan and Keohane (2006: 427), to serve for accountability purposes, the information provided "must be (a) accessible at a reasonable cost, (b) properly integrated and interpreted, and (c) directed to the accountability holders. Furthermore, (d) the accountability holders must be adequately motivated to use it properly in evaluating the performance of the relevant institutional agents" (see also Dingwerth 2007; Schmidt and Wood 2019). In this regard, ICANN made a tremendous effort to ensure the transparency of the IANA transition process, which was acknowledged even by the most critical participants and observers (Hill 2016). It provided Web sites for the IGC, the CWG-Stewardship, and the CCWG-Accountability, releasing summaries and transcripts of all meetings, as well as draft versions of the document they produced in several languages. The Web sites also contained a news section informing readers about the advancements in the process, and a section with short biographies of the working groups' members (even if many were missing or incomplete). Moreover, several mailing lists readable by the public were arranged.

The process also included numerous face-to-face meetings in different worldwide locations as well as web seminars. Thus, information about the IANA transition was undoubtedly made easily accessible for anyone interested. Paradoxically, this massive commitment to transparency produced an information overload, which can be difficult to manage.

## 6.2    Discursive Quality

### 6.2.1    Theoretical and Methodological Notes

If procedural quality relates to rules, procedures, and the institutional design of the decision-making process, the discursive quality dimension, instead, focuses on more substantial aspects of the deliberation. As seen in Chapter 2, two approaches can be employed in the analysis of the discursive quality dimension.

The first is grounded on deliberative democracy theory (Cohen 1989; Bächtiger et al. 2018; Chambers 1996), and in particular on Habermas' discourse ethics (Habermas 1999), which aims at identifying the conditions of an ideal deliberative procedure. Some of these conditions, such as universality/inclusiveness, accountability, and fairness/equality, have already been addressed. More specific conditions can be summarized in two criteria: rationality, which requires that consensus should be reached only by the means of rational argumentation, avoiding any form of coercion, and reciprocity, which refers "to the extent to which impartiality and respect are present in a given discourse and participants approach deliberations with the aim of reaching consensus" (Dingwerth 2007: 31). In Dingwerth's view, however, analyzing the dimensions of rationality and reciprocity leads to methodological difficulties, forcing reliance mostly on participants' perceptions.

For this reason, in this chapter, we adopt the set of indicators elaborated by Steenbergen et al., which can be applied to transcripts (Steenbergen et al. 2003; see also Schouten et al. 2012). These are:

1. participation (detected through the interruptions of speakers);
2. the level of justification through which participants motivate their position;
3. the reference to the common good, based both on the utilitarian (the greatest good for the greatest number of subjects) and difference principles (conception of the common good as favoring the most disadvantaged groups);
4. statements affirming respect about other groups, their demands, and counterarguments; and
5. attitudes toward constructive politics, which refers to the tendency of participants to modify their original position to reach mediation and compromise and to achieve a final agreement.

The second approach shifts the focus from the conditions of deliberation to its contents. In this view, assessing the discursive quality of a multi-stakeholder process means to look at its discourse balance (Dingwerth 2007; Schouten et al. 2012; Dryzek 2011), or in other words, it requires to take into account:

1. the number of different discourses that are represented within the decision-making process compared to the different positions that already exist in the area of concern;
2. the presence of a dominant discourse, as well as of hegemonic practices aiming at inhibiting or delegitimizing different points of view; and
3. the extent to which alternative discourses are taken into account during the discussion and in the outcome.

In this view, discourses are "shared set of concepts, categories, ideas that provide its adherents with a framework for making sense of situations, embodying judgments, and fostering capabilities" (Dryzek 2006).

To empirically detect what discourses animated the IANA transition deliberation, we first retraced the discourse coalitions that emerged in the debate on Internet governance in recent decades, as detected by previous literature and research. In doing so, we employed a detection scheme that we elaborated to investigate discourse coalitions at WSIS + 10 (Santaniello and Palladino 2017). According to Hajer (1993, 1995), a discourse coalition is a group of actors sharing a set of narratives, or story-lines, on a policy problem. Following the insights from policy narrative studies (Jones et al. 2014; Stone 2011), we have broken down discourses' narratives into a series of structural elements, namely:

1. a definition of the policy object and/or of the policy problem;
2. a public of victims damaged by the policy problem;
3. the causes of the problem and/or the subject to be blamed for causing the problem;
4. the subject supposed to solve the problem;
5. a policy solution proposed as the consequential "moral of the story."

We then analyzed the presence of these structural narrative elements within IANA transition documents and discussions to detect which discourses were produced during the decision-making process.

Both for deliberativeness and discourse balance, the corpus of analysis consisted of the transcripts of selected key points of the process. We focused on what could be considered the heart of the process, taking into account the part of the work of the CWG-Stewardship dedicated to the development of post-transition oversight and accountability arrangements. As noted, the IANA transition has been an extremely complex process, developed over two years of discussions, taking place in different working groups that were often subdivided into further entities. Performing a complete analysis of the deliberations that occurred during the IANA transition process would exceed the scope and the possibilities of this section.

Before moving to the assessment of the deliberativeness and discourse balance dimensions, the next section traces back the discussion about the post-transition oversight arrangements, highlighting the principal positions and arguments on the topic and their evolution during the debate.

### 6.2.2    The Discussion on Post-transition Oversight Arrangements

To summarize a very complex debate, it could be said that the CWG deliberation on the replacement of the NTIA's role developed around two alternative positions, later called the "internal" and the "external" models.[8] The internal model, fostered mainly by registries and the technical community, was based on the assumption that the NTIA oversight role was merely an administrative or "clerical" function.[9] In this view, the NTIA's role was simply the verification that IANA functions

---

[8] The two positions were raised at an early stage of the debate. In particular, refer to meetings 1, 13, and 14 of the RF3, and meetings 11, 18, 20, 21, 23, 31, and 32 of the CWG-Stewardship. All the transcripts are available at https://community.icann.org/display/gnsocwgdtstwrdshp/Meetings (Last accessed December 2019).

[9] The NTIA oversight has been defined "clerical" or "administrative" in reference both to the performance review and the authorization of the delegation or re-delegation process (the process assigning or re-assigning a TLD to a manager and inscribing it in the root file). For example, the NTIA role has been described as a "technical administrative policy review" (SSAC member, RFP3 Meeting 1); an "administrative task involving some technical issues, some technical checks and so forth" (GNSO member, CWG Meeting 4); and

were performed correctly, according to the procedures and service level requirements established in the contract and related documents, as well as to the policies autonomously developed by ICANN, the IETF, and the RIRs in a bottom-up manner. Further, the supporters of the internal model argued that NTIA verifications had a clerical nature (requests for the due documentation such as performance reports, audit data, and reports), while IANA customers are the de facto evaluators of the IANA technical performance. As a consequence, from this perspective, "the stewardship just should be transferred to the ICANN community,"[10] and the oversight should be ensured through internal mechanisms.[11] Moreover, in this view, a special role should be assigned to registry operators since they are the direct customers of IANA services and have the necessary expertise and economic reasons for evaluating the adequacy of the IANA operator's performance.

By contrast, the "external model" solution was supported mostly by non-technical actors, such as governments, commercial and non-commercial constituencies, and academics, who did not agree with a merely technical interpretation of NTIA oversight. They argued that the NTIA contract, being subject to renewal and foreseeing the possibility of awarding the IANA functions to a different contractor, had been the most powerful mechanism to influence "ICANN conduct and its accountability"[12] by engaging it in a "legal binding commitment."[13] Thanks to that contract, NTIA could realize in-depth periodic performance reviews and set the "operational requirement for the IANA function,"[14] which are processes with a reach far beyond the technical dimension. Actors advancing the external model pointed out that purely internal mechanisms

"a very clerical and notarial function" (ICANN Staff, CWG Meeting 12). In particular, the issue was investigated during CWG meetings 7 and 8.

[10] ALAC member, RFP 3 Meeting 1.

[11] "I think ICANN is the steward at that point to the extent that they remain in. And the job of us and the accountability group together is to make sure that that is appropriately and adequately an accountable stewardship" (GNSO member, CWG Meeting 32).

[12] GNSO, Non-Commercial User Constituency member, CWG Meeting 11.

[13] ISP member, CWG Meeting 7. See also discussion held in RFP3B Meeting 1.

[14] GNSO member, RFP3 Meeting 1.

would have left the ICANN board unconstrained.[15] Further, they argued that, in the better case, this system could ensure accountability toward the ICANN community but not toward the broader global multistakeholder community.[16] According to them, this "separability" mechanism, ensured by the possibility of the NTIA to rebid the IANA functions, had to be preserved, replacing the role once performed by the NTIA with an independent multistakeholder legal entity outside ICANN.

It is worth noting that both sides of this contention agreed on the principle of "separability" (even if it encountered some resistance among internal model supporters), and both considered it an extreme remedy, a "nuclear option" for a "doomsday scenario"; however, they interpreted it in two different ways.[17] External model advocates considered separability as the possibility that an independent entity would remove IANA functions from ICANN and reassign them to another subject. Those supporting the internal model referred to separability in terms of functional separation between policy development and IANA operations activities, where only the operations could be reassigned to another IANA function operator and stewardship and policy-making would remain within ICANN.

Similarly, both the sides agreed on principles such as "do no harm" and "if it ain't broke, don't fix it," according to which the transition process had to remain limited to the minimal changes needed to prevent endangering the functioning of the DNS. However, even in this case, the principles were interpreted in different ways by the two factions. For internal model supporters, "do no harm" meant avoiding creating novel structures and keeping changes within ICANN, a well-known organization capable of solving potential future problems through intercommunity dynamics. For the external model advocates, the commitment to minimal changes implied maintaining, as far as possible, the existing mechanisms and only replacing the NTIA with a multistakeholder body.

The first draft proposal released on December 1, 2014, achieved a compromise between the two models, breaking down the oversight role

---

[15] The point was particularly clearly by several ccNSO, civil society, and academic members during CWG Meeting 32.

[16] The argument was raised mainly by GAC and ccTLD representatives during CWG Meeting 7.

[17] Conceptions of separability were widely debated during the 7th and 11th CWG meetings.

of the NTIA into more specific tasks and assigning them to different entities. The contracting function was entrusted to a new independent legal entity, provisionally called Contract Co, conceived as a lightweight organization with little or no staff and as a vehicle to enforce the provision of the contract with the IANA function operator, following the instructions of another organization, the multistakeholder review team (MRT). The MRT would have included formally selected representatives from all of the relevant communities (even if the exact composition remained unspecified), conducting periodic performance reviews and making decisions for Contract Co. (including the definition of IANA contract terms and whether or not the bidding process was re-opened). Another body composed of representatives of registry operators, the Customer Standing Committee (CSC), would have received and reviewed the day-to-day performance reports of the IANA function operator, collaborating with the MTR for the definition of service level and performance indicators for the IANA naming function. Several arrangements were put in place to replace the NTIA's authorization role, including the publication of all requests for changes to the root zone file, an independent certification for delegation and re-delegation requests, and the creation of an independent appeal panel.

The public comment period re-opened the discussion about the internal/external model. Many relevant stakeholders, such as Google, the business constituency, Nominet, SDIN, the ALAC, and ICANN itself, contested the Contract Co. instrument, raising concerns about the cost of the new structure and the risks for the DNS related to litigation, controversies around its jurisdiction, and the possibility of capture by special interests (in particular, the nationalization by the country under which Contract Co. would have been incorporated). These players asked the CWG to consider and develop alternative proposals capable of matching the principle of separability through internal arrangements, before making a final decision.

The CWG welcomed the suggestions and, during the ICANN Istanbul meeting of March 2015, decided to abandon the external model. This choice was justified (mostly by registries' representatives) arguing that Contract Co. was an "unproven" and "indefinite" entity putting the DNS at greater risk than an internal solution, as regards the stability, security, and continuity of the DNS. Other actors replied to this claim affirming that any solution that removed the NTIA's role would have entailed some risks and uncertainty.

Nevertheless, the most decisive motivation for the turn toward an internal solution was the concern that the NTIA, or the US Congress, could not accept an external solution involving an independent entity whose jurisdiction could be placed outside the USA. Warnings about possible reactions by the NTIA had been raised since the first stages of the CWG deliberation.[18] They entered the agenda of the CWG after remarks made by Lawrence E. Strickling, Assistant Secretary for Communications and Information at the Department of Commerce and Administrator of the NTIA, during the State of the Net Conference on January 27, 2015.[19] Strickling posed questions about the risks related to the establishment of new entities and asked if alternative options had been adequately considered. The prospect of failing to receive NTIA approval was sufficient to persuade some CWG members to give up on the idea of an independent, external contracting entity.

A second draft proposal provided that the NTIA would transfer the IANA functions, related resources, and the rights for contracting the IANA functions directly to ICANN. In turn, ICANN would contract a new legal entity, the Post-Transition IANA (PTI), established as an affiliate nonprofit corporation controlled by ICANN, granting the PTI the rights and obligations to serve as the IANA functions operator. At the same time, a series of novel internal mechanisms (such as the IANA Function Review and the CSC) would be established to oversee the performance of the IANA function operator. These mechanisms could eventually give rise to a separation process leading PTI to cease the performance of the IANA naming function.

### 6.2.3    Evaluating Deliberativeness and Discourse Balance

How close was the discussion leading to the final proposal to an ideal deliberative procedure? Let us take into account the elements underlined by Steenbergen et al. (2003): participation, justification, reference to the common good, respect, and attitude to constructive politics:

---

[18] See CWG Meetings 7 and 9 transcriptions.

[19] https://www.ntia.doc.gov/speechtestimony/2015/remarks-assistant-secretary-strickling-state-net-conference-1272015. The Strickling statements was one of the points in the agenda of the 14th meeting of the CWG-RFP3 subgroup on February 2, 2015. They were also crucial in the 32nd CWG Meeting discussions, where the internal model prevailed.

1. As regards participation, we observed that all the speakers could express their views without restriction or obstacles; no one was deliberately interrupted. In a broader sense, however, we notice that only a small number of CWG members and participants regularly attended the meetings (considering both live and remote participation),[20] and even fewer took part actively in the discussion.

2. The level of justification through which participants motivated their positions was generally sophisticated. Speakers made a continuous effort to explain their reasons and the rationale for their stances. Participants generally built their arguments on the analysis of the current mechanisms and disagreed about the possible costs and benefits of the proposed change, discussing and comparing alternative solutions. Nevertheless, it is worth noting that this rational and argumentative approach was perverted by considerations about NTIA preferences, which moved the discussion from the search for the best institutional setting to solutions with a better chance of being approved by the NTIA.

3. References to a utilitarian conception of the "common good" could be found in the shared and numerous concerns about the stability, continuity, and security of the DNS, which could be conceived as commitments toward the well-being of the entire Internet community. By contrast, conceptions of the common good according to the difference principle (in favor of the most disadvantaged groups) were almost absent. Most of the time, stakeholders justified their positions by calling into question their narrow and particular interests. For example, registries supported an internal oversight model, limited to service levels in which they would have played a leading role as direct customers of IANA services. Similarly, ccTLD registries and operators advocated the lightest possible form of oversight, without interfering with ccTLD policies developed at the national level. Governments claimed a meaningful and effective role in future arrangements.

4. The discussion was generally respectful. Participants often valued the presence and contributions of other stakeholders and the proposals advanced by each group were usually recognized as legitimate by

---

[20] Please refer to attendance data at https://community.icann.org/display/gnsocwgdt stwrdshp/Attendance+Log+CWG-Stewardship. Accessed 12 June 2020.

others. Assessing the degree of respect toward others' counterargument is much more difficult. According to Steenbergen et al. (2003), this implies that counterarguments are not ignored and then that they are valued. In this regard, the IANA transition deliberation was extremely variegated. Some counterarguments went unconsidered and seemed to be lost in the flow of discussion. Sometimes they were undermined by opposing participants, while at other times they were welcomed and brought to a meaningful and generally agreed change.

5. Participants in the deliberation showed a great attitude toward constructive politics. Almost all of the members were able to review their original positions, or at least to cease their objections, to allow the process to move forward. People taking part in the discussion were generally animated by a strong will to realize the transition, particularly to remove US oversight. Several statements indicate that this objective took priority over any particular preferences concerning the arrangements through which it would be carried out.

Regarding discourse balance, we check to what extent the IANA transition was a pluralistic process representing the constellation of discourses that exist in the public sphere on Internet governance. Discourse balance is crucial if one wants to seriously consider the legitimacy of the forms of deliberative democracy that are supposed to take place within transnational networks and multistakeholder initiatives. Without such discursive pluralism, a transnational decision-making process could still be rational and consensual, but would not comply with democratic legitimacy criteria since not all the most relevant ideas, opinions, and preferences of those subjected to the resulting decision were represented and discussed during the deliberation.

Based on the literature and previous research (Santaniello 2016; Santaniello and Palladino 2017; Mueller 2010; Chenou 2014; Hofmann 2007; Drake 2004), we identified three main discourses that animated the debate around Internet governance and DNS since the 1990s:

1. *Neoliberal discourse.* The US government has fostered this discourse, gathering around it the most of the other Western countries, the private sector, and the technical community (at least, after the establishment of ICANN). In this view, the Internet, and notably, the

DNS, has been defined as a "networking technology" with a "commercial value" (USG-DoC 1998), a "global facility,"[21] or as "technical standards developed in an open, market-driven paradigm."[22] In this discourse, policy problems are conceived in terms of the diffusion, development, and efficiency of the Internet, while its political and social implications tend to be ignored or denied. The idea at the core of the neoliberal discourse is that "market mechanisms that support competition and consumer choice should drive the technical management of the Internet" (USG-DoC 1998) since they ensure innovation, openness, and the necessary flexibility to provide rapid responses to technological change and user needs. State-centric models of governance, with their bureaucracy, slow pace, and political compromise, are considered a threat to the functioning and evolution of the Internet. Consequently, private operators are seen as leading actors.

2. *Sovereigntist discourse.* This conceives the Internet as a national strategic asset that states have to control to guarantee the well-being of their populations and the national interest. Policy problems are extended far beyond technical functionality to include a wide range of public issues related to the Internet. This narrative criticizes the unregulated global expansion of ICT, driven by market forces, for not being able to face the negative externalities it produces, particularly for the growing inequalities and asymmetries among states and populations in the distribution of opportunities and threats from the new technologies. Thus, this coalition, led by BRICs (Brazil, Russia, India, and China)[23] and followed by other developing countries, proposes the regulation of the Internet through national legislation and multilateral agreements. Concerning DNS management, sovereigntists call for greater participation of all governments "on an equal footing" and criticize the special role of the US government.

---

[21] USG submission to WGIG, 2004, https://www.wgig.org/docs/usa.doc (Last accessed 16 March 2020).

[22] IEEE submission to WSIS + 10, 2015, https://publicadministration.un.org/wsis10/ (Last accessed 10 March 2020).

[23] However, since Netmundial and the approval of the Marco Civil, Brazil moved toward a constitutional discourse. Its approach to Internet governance mixes state regulation, multistakeholder governance, and effective commitment to human rights.

3. *Constitutional discourse.* This third discourse coalition represents a set of concerns shared mainly among IGOs (like UNESCO) and civil society associations. Chenou (2014) called it "the global public good discourse" while we refer to it as the "Constitutional Coalition" (Santaniello and Palladino 2017). For actors sharing and advancing this discourse, the "Internet is a global public good, which must be managed in the interests of all the world's peoples."[24] Unlike sovereigntists, the acknowledgment of the public nature of the Internet is not functional to the legitimization of some public authority. Instead, it aims to stress the urgency of strengthening the people's control over technological development and the respect of net-citizens' rights. They identify the main problems of Internet governance in "a lack of democracy; an absence of legitimacy, accountability, and transparency; excessive corporate influence and regulatory capture; and too few opportunities for effective participation by people,"[25] blaming both governments and private actors for systematic human rights violations.

An analysis of the statements and contributions submitted during the preparatory phase confirmed that the debate on the IANA transition included these three discourses even before it started. On 48 documents analyzed, 19 included one or more structural elements of the neoliberal discourse; 16 of the sovereigntist discourse; and 6 of the constitutional discourse.

However, the deliberation that occurred within the CWG-Stewardship seems to have been entirely inscribed within a discursive neoliberal framework, while references to the other two discourses were much more marginal. Even the two competing positions about the internal/external model shared several assumptions entirely consistent with the neoliberal discourse. In particular, despite the differences in the NTIA's role, most of the participants in the CWG-Stewardship discussion agreed on a conception of the IANA functions as a merely technical and administrative

---

[24] JustNet and APC submission to WSIS + 10, 2015, https://publicadministration.un. org/wsis10/ (Last accessed 10 March 2020).

[25] Delhi Declaration, available at https://justnetcoalition.org/delhi-declaration (Last Accessed 16 March 2020).

service.[26] This is not surprising if we consider that it was a widely shared belief among the founding fathers of the Internet that "issues concerning the constitution and management of the underlying [Internet] infrastructure should be insulated to the greatest degree possible from potentially troublesome policy discussions" (Drake 2004: 127).

These conceptions became key points of the neoliberal discourse, and they were embedded in NTIA/ICANN language and documentation as a market-oriented paradigm. This discourse portrayed IANA functions as technical and administrative services provided to customers within a market-driven registry system, enabling the separation between policy development and operational activities, and nurturing the assimilation of the DNS management to the domain of private governance. Ultimately, this narrow technical frame prevented the acknowledgment of the public good nature of the IANA functions, and, even more, of their essence as public policy issues.

This definitional strategy reflected widespread concerns about DNS capture by governments, or otherwise its cooptation "for purposes other than those for which they were initially designed" (DeNardis and Musiani 2016: 5). On the other hand, it is widely recognized that the "Internet's technical protocols and standards embody and affect non-technical norms and values, with the DNS often used as a prime illustration of the principle" (Post and Kehl 2015: 20). However, defining a space both neutral and technical does not solve its policy issues; rather, it is a way to conceal both problems and solutions.

Discursively, this agreement on the neoliberal definition of the IANA functions is fully consistent with the internal oversight model, which was defined as a service performance review. The neoliberal discourse also made it natural to have the operational communities as the stewards of the IANA functions and the registries as "special" stakeholders in their role of "direct" customers. By contrast, the external model position did not show the same degree of consequentiality with the assumption about the merely technical and private nature of the IANA functions. Rather, such a model would have benefited from the acknowledgment of the IANA functions as a public good. In this latter case, it would have been easier

---

[26] Please refer to conversations held during RFP3b meeting on January 14, 2014, where many participants of different constituencies agreed in defining IANA functions as "clerical," "administrative," "technical," or "operational." https://community.icann.org/pages/viewpage.action?pageId=51417868 (Last Accessed 16 March 2020).

to justify the establishment of an independent multistakeholder body as a guarantor of the public interest.

The discrepancy between the discourses detected in the public sphere during the preparatory phase and those developed during the IANA transition deliberation demonstrates the hegemony nature of the neoliberal discourse during the deliberative process, which inhibited references to the structural elements of the other two discourses, and in particular to the public good nature of the IANA functions.

The process relied on pre-existing organizations (ICANN, the IETF, and the RIRS) fully integrating the neoliberal discourse within their culture and practices. Moreover, most members and participants at the CWG-Stewardship had been socialized to the ICANN system, belonging to one of its structures or attending its meetings. Even if this does not mean that they necessarily embraced a neoliberal view, it nevertheless suggests that they were aware of the boundaries of acceptable ideas and arguments within that context, also in consideration that the ICANN board, the NTIA, and the US Congress had the power to approve or reject the transition plan.

To conclude, the long-standing neoliberal plan of the US government and the NTIA to "privatize" the DNS placed the IANA transition within a precise system of definitions, concepts, references, and assumptions that constrained the development of alternative policy discourses and limited the political action of sovereigntist and constitutional coalitions. The analysis we conducted suggests some considerations among the relationship between deliberativeness and discourse balance. A shared discourse could stabilize a network or multistakeholder body, making mutual understanding easier. This is also true for deliberativeness. Common assumptions reduce the possibility of intractable conflicts and facilitate a rational discussion. Nevertheless, as Dryzek (2011) pointed out, this greater stability and practicality is bought at a cost:

> Domination of a network by a single discourse means that alternative interpretations that can only be generated within other discourses are not heard [...] the sorts of differences and challenges that are grist for a deliberative conception of democracy get lost. [...] If a network shuts out contending discourses, then it risks becoming progressively illegitimate with time, as well as ineffective in problem-solving terms. (Dryzek 2011: 129–130)

# References

Bächtiger, A., Dryzek, J. S., Mansbridge, J., & Warren, M. (Eds.). (2018). *The Oxford Handbook of Deliberative Democracy*. Oxford: Oxford University Press.

Beisheim, M., & Dingwerth, K. (2008, June). *Procedural Legitimacy and Private Transnational Governance: Are the Good Ones Doing Better?* Report, SFB Research Center, Freie Universität Berlin, DE. https://www.sfb-gov ernance.de/en/publikationen/sfb-700-working_papers/wp14/SFB-Govern ance-Working-Paper-14.pdf. Accessed March 17, 2020.

Buchanan, A., & Keohane, R. O. (2006). The Legitimacy of Global Governance Institutions. *Ethics & International Affairs, 20*(4), 405–437.

Chambers, S. (1996). *Reasonable Democracy: Jürgen Habermas and the Politics of Discourse*. Ithaca: Cornell University Press.

Chenou, J. M. (2014). From Cyber-Libertarianism to Neoliberalism: Internet Exceptionalism, Multi-stakeholderism, and the Institutionalisation of Internet Governance in the 1990s. *Globalizations, 11*(2), 205–223.

Cohen, J. (1989). Deliberation and Democratic Legitimacy. In A. P. Hamlin & P. Petitt (Eds.), *The Good Polity: Normative Analysis of the State* (pp. 18–34). Oxford: Wiley.

Dellmuth, L., Scholte, J., & Tallberg, J. (2019). Institutional Sources of Legitimacy for International Organisations: Beyond Procedure Versus Performance. *Review of International Studies, 45*(4), 627–646. https://doi.org/10.1017/S026021051900007X.

DeNardis, L., & Musiani, F. (2016). Governance by Infrastrucure. In F. Musiani, D. L. Cogburn, L. DeNardis & N. S. Levinson (Eds.), *The Turn to Infrastructure in Internet Governance* (pp. 3–21). New York: Palgrave Macmillan.

Dingwerth, K. (2007). *The New Transnationalism, Transnational Governance and Democratic Legitimacy*. New York: Palgrave Macmillan.

Doherty, D., & Wolak, J. (2012). When Do the Ends Justify the Means? Evaluating Procedural Fairness. *Political Behavior, 34*, 301–323.

Drake, W. J. (2004). Reframing Internet Governance Discourse: Fifteen Baseline Propositions. In D. MacLean (Ed.), *Internet Governance: A Grand Collaboration* (pp. 122–126). New York: UN ICT TF.

Dryzek, J. S. (2006). *Deliberative Global Politics: Discourse and Democracy in a Divided World*. Cambridge: Polity Press.

Dryzek, J. S. (2011). *Foundations and Frontiers of Deliberative Governance*. Oxford: Oxford University Press.

Habermas, J. (1999). *Between Facts and Norms: Contributions to a Discourse Theory of Law and Democracy*. Cambridge: MIT Press.

Hajer, M. A. (1993). Discourse Coalitions and the Institutionalisation of Practise. In F. Fischer & J. Forester (Eds.), *The Argumentative Turn in Policy Analysis and Planning* (pp. 43–76). Durham and London: Duke University Press.

Hajer, M. A. (1995). *The Politics of Environmentalism*. Oxford: Oxford University Press.

Hill, R. (2016). Internet Governance, Multi-Stakeholder Models, and the IANA Transition: Shining Example or Dark Side? *Journal of Cyber Policy*. https://doi.org/10.1080/23738871.2016.1227866.

Hofmann, J. (2007). Internet Governance: A Regulative Idea in Flux. In R. K. J. Bandamutha (Ed.), *Internet Governance: An Introduction* (pp. 74–108). Icfai: University Press.

Hooker, B. (2005). Fairness. *Ethical Theory and Moral Practice, 8*(4), 1–24.

Jones, M. D., Shanahan, E. A., & McBeth, M. K. (2014). *The Science of Stories: Applications of the Narrative Policy Framework in Public Policy Analysis*. New York, NY: Palgrave Macmillan.

Mueller, M. (2009). ICANN Inc.: Accountability and Participation in the Governance of Critical Internet Resource. *The Korean Journal of Policy Studies, 24*(2), 91–116.

Mueller, L. M. (2010). *Networks and States: The Global Politics of Internet Governance*. Cambridge: MIT Press.

Post, D. G., & Kehl, D. (2015). *Controlling Internet Infrastructure: The 'IANA Transition' and Why It Matters for the Future of the Internet*. Part 1. https://static.newamerica.org/attachments/2964-controlling-internet-infrastructure/IANA_Paper_No_1_Final.32d31198a3da4e0d859f989306f6d480.pdf. Accessed September 15, 2018.

Risse, T. (2006). Transnational Governance and Legitimacy. In Arthur Benz & Yannis Papadopoulos (Eds.), *Governance and Democracy Comparing National, European and International Experiences* (pp. 179–199). New York: Routledge.

Santaniello, M. (2016). Net democracy: la sfida democratica all'Internet governance. In E. De Blasio & M. Sorice (Eds), *Innovazione democratica. Un'introduzione* (pp. 63–86). Roma: Luiss University Press.

Santaniello, M., & Palladino, N. (2017). *Shaping Words to Shape Policy Process: Discourse Coalitions in the Internet Governance Ecosystem*. Paper Presented at the 1st GIG-ARTS Conference, Paris, March 30, 2017.

Schmidt, V., & Wood, M. (2019). Conceptualizing Throughput Legitimacy: Procedural Mechanisms of Accountability, Transparency, Inclusiveness and Openness in EU Governance. *Public Administration*. https://doi.org/10.1111/padm.12615.

Scholte, J. A., & Tallberg, J. (2018). Theorizing the Institutional Sources of Global Governance Legitimacy. In J. Tallberg, K. Bäckstrand, & J. A. Scholte (Eds.), *Legitimacy in Global Governance: Sources, Processes, and Consequences* (pp. 56–74). Oxford: Oxford University Press.

Schouten, G., Leroy, P., & Glasbergen, P. (2012). On the Deliberative Capacity of Private Multistakeholder Governance: The Roundtables on Responsible Soy and Sustainable Palm Oil. *Ecological Economics, 83*, 42–50.

Steenbergen, M. R., Bachtiger, A., Sporndli, M., & Steiner, J. (2003). Measuring Political De-liberation: A Discourse Quality Index. *Comparative European Politics, 1*, 21–48.

Stone, D. (2011). *Policy Paradox: The Art of Political Decision Making* (3rd ed.). New York: Norton.

USG-DoC. (1998). *Green Paper*. https://www.ntia.doc.gov/federal-register-notice/1998/statement-policy-management-internet-names-and-addresses. Accessed March 16, 2020.

# CHAPTER 7

# The Output Legitimacy of the IANA Transition Process

**Abstract** The chapter assesses the output legitimacy of the IANA transition from a normative perspective. In detail, the chapter investigates whether the outputs of the process effectively achieved the intended objective, namely the transfer of the IANA functions stewardship to the Internet global multistakeholder community. Furthermore, the chapter assesses whether the new arrangements and the bylaw's modifications made it possible to remedy the well-known limitations of ICANN governance, both in terms of institutional and outcome effectiveness.

**Keywords** IANA transition · Output legitimacy · ICANN · Bylaw · Accountability · .org

## 7.1   Output Legitimacy
## in Constituent Policy-Making

Multistakeholderism is based on a precise expectation for what concerns the output of a decision-making process; that is, to ensure "outcomes which promise to settle the issue in the long run [...] impossible to achieve by less diverse constellations of actors" (Hofmann 2016: 32). This expectation represented the main rationale of the IANA transition process

© The Author(s) 2021    127
N. Palladino and M. Santaniello, *Legitimacy, Power, and Inequalities in the Multistakeholder Internet Governance,*
Information Technology and Global Governance,
https://doi.org/10.1007/978-3-030-56131-4_7

and provided it with a clear objective, that is, on a more general level, to "support and enhance the multistakeholder model of Internet policymaking and governance," and, on an operational level, to "transition the stewardship of key Internet functions to the global multistakeholder community" (NTIA 2014). However, because of the ambiguity and diverging interpretations of concepts such as "multistakeholder community" or "Internet community," there is no unique and impartial way to set up effective criteria to assess whether and to what extent the declared goal of the IANA transition was achieved. To develop a clearer picture of the transition's results, we rely on Bäckstrand's distinction between "outcome effectiveness," which refers to the actual achievement of concrete goals, and "institutional effectiveness," which takes into account the appropriateness of the institutional design established by a governance structure to reach its objectives (Bäckstrand 2006). Here, we consider outcome effectiveness and institutional effectiveness as two analytical dimensions of the IANA transition's output legitimacy.

Outcome effectiveness helps us to evaluate the transition against the substantial aim of the IANA transition, which was to establish a more legitimate authority over IANA functions and DNS governance by moving the IANA stewardship from the NTIA to the "global multistakeholder community." Outcome effectiveness is often operationalized into quantitative measures to evaluate the goals of a policy process. However, in the case of the IANA transition, this approach would merely turn into a checklist concerning the timing of the proposal's delivery and the approval or rejection of the transition by US authorities. Instead, we use the concept of outcome effectiveness more broadly, including the long-term effects of the transition on conflicts around DNS management. Hence, our analysis of this dimension takes into account stakeholders' behavior toward wider aspects of ICANN governance and IANA functions, as this has emerged over a longer period after the transition. This analysis is guided by the following research question: Have the new arrangements appeased traditional conflicts and tensions related to the DNS and ICANN governance structure, or have these disputes re-emerged after the 2016 transition? To answer this question, we discuss two crucial cases: (I) the so-called Internet sovereign law approved in 2019 in the Russian Federation and (II) the attempt by ISOC to sell the management of the .org TLD to a private equity firm.

Concerning the second dimension of output legitimacy, institutional effectiveness, in Chapter 2 we already noted that, in constituent policies with outputs consisting of new institutions or mechanisms, the output of the policy process produces the structural conditions that affect both the input and throughput dimensions of the novel setting. This is very much the case for the IANA transition, whose objective was to redesign the governance arrangements for the IANA functions to replace NTIA oversight. Here, by institutional effectiveness we mean the capability of institutional design produced by the constituent policy-making to empower the Internet multistakeholder community in the management of the IANA functions. Thus, to assess the institutional effectiveness of the transition process, we adopt some criteria that already guided our analysis of its input and throughput legitimacy. While some criteria previously used in our analysis, such as representativeness, fairness, and discursive quality, are dependent on contingent policy processes and can be assessed only by taking into account a specific set of practices, other criteria, such as inclusiveness, balanced representation, and accountability, can be used to evaluate the structural features of the post-transition governance arrangement. As a consequence, our analysis focuses on ICANN's post-transition bylaws, which we compare with the previous bylaws based on the inclusiveness, stakeholder balance, and accountability criteria. The research question we seek to answer is: Is the new institutional arrangement significantly more inclusive, balanced, and accountable than the previous one?

## 7.2   INSTITUTIONAL EFFECTIVENESS: THE NEW ICANN BYLAWS

The first post-transition ICANN bylaws came into effect on October 1, 2016 (ICANN 2016). The main novelties of these bylaws are represented by the establishment of new bodies and mechanisms aimed at redesigning ICANN governance after the end of NTIA oversight, namely the Empowered Community (EC), the Public Technical Identifiers (PTI), the Customer Standing Committee (CSC), and a set of new review processes known as the IANA naming functions reviews (IFRs).

The EC was established by Article 6 of the ICANN bylaws as a "non-profit association formed under the laws of the State of California," particularly the California Corporations Code. According to the post-transition bylaws, the EC shall not have any directors, officers, or employees. Also, it must be based in the same location as the principal office of ICANN. It consists of five collective "Decisional Participants," namely the three ICANN supporting organizations (the ASO, the ccNSO, and the GNSO) and two advisory committees, the ALAC and the GC. The Decisional Participants act through their respective chairs or other delegated persons (collectively named the "EC Administration"), who receive from their respective organizations a binding mandate on each decision to be made. The EC can exercise its powers and rights through a set of procedures that are detailed in Annex D of the new bylaws. These rights and powers may be conceptually grouped into four categories.

The first group concerns the so-called approval actions that relate to the power to approve fundamental bylaw amendments, articles of incorporation amendments, and asset sales proposed by the board. The approval process begins with a forum convened by the EC Administration, the so-called Approval Action Community Forum. After the discussion held in the forum, each Decisional Participant has to inform the EC Administration in writing about its position on the proposal. The action is approved if it is supported by three or more Decisional Participants and not objected to by more than one Decisional Participant.

The second group of EC powers consists of "rejection actions," which can be initiated against (i) ICANN budgets and standard bylaw amendments; (ii) IANA budgets, operating plans, and strategic plans; and iii) actions aimed at modifying the governance structure of the PTI (PTI Governance Actions). These types of action can be started following a petition, which can be presented by any individual to a single Decisional Participant and has to be supported by four or more Decisional Participants (three in the case of rejection actions relating to a standard bylaw amendment), and not objected to by more than one Decisional Participant.

The third group of EC rights and powers relates to the possibility of removing a single member of the ICANN Board of Directors or to recall the entire board, other than the president. In the case the action is exercised against a single director, an individual petition may trigger the process. If the nominating committee appointed the concerned director, the petition might be addressed to a particular Decisional Participant and followed by a community forum. After this process, if three or more Decisional Participants support the removal action and no more than one Decisional Participant objects, the concerned director is removed. If the concerned director was nominated by a supporting organization or an advisory committee, the individual petition has to be addressed to the relevant Decisional Participant. The petition must be supported by a supermajority (usually a three-quarters majority) as determined by the internal procedures of the applicable Decisional Participant. In this case, a community forum (plus a public comment period) also follows the petition, which is approved if it gains support from three or more Decisional Participants and no more than one Decisional Participant objects.

The mechanism concerning the recall of the entire board can be started by an individual petition to one Decisional Participant. The petition has to be supported by at least two Decisional Participants, and it is followed by a community forum for public discussion. The recall of the board is approved if the petition is supported by four or more Decisional Participants and is not objected to by more than one Decisional Participant.

The fourth group of EC powers and rights includes a set of reviews about ICANN processes and board decisions. Some of these reviews were already in use before the transition, such as the Reconsideration Request and the Independent Review Process. The EC gained the power to initiate these, just like any other person or entity materially affected by an action or inaction of the ICANN board or staff. Furthermore, the new bylaws gave the EC an important role concerning the two newly established review processes: IFRs and the IANA Naming Function Separation Process.

The IFRs consist of reviews of the PTI's performance in the IANA naming functions. These reviews are categorized into two groups: regular periodic reviews and special reviews. Each review is carried out by an ad hoc IANA function review team (IFRT), which is not a standing body and is automatically disbanded at the end of the reviewing process. Each IFRT is composed of three representatives appointed by the ccNSO and five appointed by the GNSO (containing two representatives appointed by the Registries Stakeholder Group, one by the Registrars Stakeholder Group, one by the Commercial Stakeholder Group, and one by the Non-Commercial Stakeholder Group). The GAC, the SSAC, the RSSAC, and the ALAC appoint one representative each. The CSC, the ASO, and the IAB also each appoint one liaison to the team. Each IFRT has two co-chairs, one selected by the GNSO among the representatives of its constituencies and one by the ccNSO among its appointed members. The recommendations advanced by the IFRTs can be approved or rejected by the ICANN board, and, in both cases, they can be contested by the EC through an approval or rejection action petition.

Another relevant novelty introduced by the post-transition bylaws is the so-called IANA Naming Function Separation Process, governed by Article 19. The separation process is the process that may lead to the termination of the contract between ICANN and the PTI and the selection of a new IANA naming function operator. This process can be triggered by a recommendation from an IFRT upon the conclusion of a periodic or special IFR. This recommendation consists of a request for the creation of a Separation Cross-Community Working Group (SCWG) to lead the process. The SCWG is established by the ICANN board only if it is approved by a supermajority of both the ccNSO and the GNSO, and if the EC has not rejected the board's approval. The composition of each SCWG is more or less that of the IFRTs, the main difference being that, in the SCWG case, the Registries Stakeholder Group of the GNSO appoints three representatives instead of two. An SCWG recommendation is implemented if it is approved by a supermajority of both the ccNSO Council and the GNSO, if it is approved by the board after a public comment period, and if the EC does not reject the recommendation.

As we have already seen, the new ICANN bylaws established a new body to replace the IANA. This new body is the Post-Transition IANA, officially incorporated as the PTI. The PTI was established as a California nonprofit public benefit corporation which, even though it was incorporated as a separate legal entity, has ICANN as its sole member. The PTI provides IANA naming functions to ICANN on the base of a renewed IANA naming function contract. The PTI works according to its own articles of incorporation and bylaws. These documents envisage a PTI board consisting of five directors: three, including the president, are persons employed by ICANN or by the PTI and who are nominated by ICANN itself, and two who are not employed by either ICANN or the PTI, who are nominated by ICANN's nominating committee.

To monitor the PTI's performance of the naming functions, the post-transition ICANN bylaws also envisaged the establishment of the CSC, which acts as a body representing ICANN's direct customers, namely top-level domain registry operators as well as root server operators and operators of other non-root zone functions. The CSC consists of two members representing gTLD registry operators, appointed by the Registries Stakeholder Group of the GNSO; two members representing ccTLD registry operators, appointed by the ccNSO; and one liaison appointed by the PTI. Other liaisons may be appointed by other supporting organizations and advisory committees. For completeness, it should be said that both the IETF and the NRO, on behalf of the protocol parameter and the numbers communities, renewed their contracts with ICANN for the operation of their respective registries, allowing ICANN to subcontract the PTI. Both the IETF and the NRO established their own service review committees to monitor compliance with the respective service level agreements. The current arrangements maintain the possibility for the operational communities to break the agreements and find a different IANA function operator, which could put at risk the consistency of the DNS.

What do these institutional innovations tell us about the levels of inclusiveness, balanced representation, and accountability of the post-transition ICANN? As for the inclusion of different interests into the new institutional design of ICANN, it is clear enough that the composition of the newly established bodies is heavily dependent upon the usual ICANN constituencies. To the end, no new interests were included in the post-transition ICANN and, as a consequence, ICANN's inclusiveness was not affected by the transition. As for balanced participation, if we look at the composition of the PTI and the CSC, we find that the representatives of ICANN's contractual partners, namely registries and registrars, still play a dominant role. In the case of the EC, the novelty in terms of balanced representation is the presence of the GAC as a Decisional Participant with voting power, while in the previous ICANN governance structure the GAC had only a consultative role with no voting members appointed to the board or the nominating committee. This shows the strengthening of a specific class of actors, national governments. On the other hand, one should consider that both the process through which the EC can exercise its powers and rights, which requires the support of three or even four Decisional Participants, and the heterogeneity of the GAC itself, does not allow discussion about extended balanced representation in ICANN.

Summarizing the institutional output of the transition, ICANN identified its preexisting constituencies and contracted partners as the "global multistakeholder community," and transferred to itself the stewardship of IANA functions. Hence, no relevant change occurred in terms of inclusiveness and the balanced representation of different interests, except for the fact that governments continue to gradually strengthen their position in ICANN.

As regards accountability, a first comment concerns the role of the EC. The EC constitutes the main internal mechanism that is supposed to replace external NTIA oversight and hold the ICANN board accountable. Within the "state analogy"[1] that inspired the framework of the Enhancing Accountability process, the EC represents the body by which the "political community" can take action against the "executive," represented by the ICANN board, in the case its behavior endangers the interest of the community itself.

Nevertheless, a complicated escalation and threshold system make these powers difficult to exercise effectively and an agreement within and between Decisional Participants hard to reach, especially on proposals coming from the ALAC and the GAC, which are the most heterogeneous advisory committees. Also, it is worth noting that, except for the outcomes of an Independent Review Process, all the decisions made by the EC (approval actions, rejection actions, removal of a director, recall of the board, re-reviews of IFRs, and SCWG recommendations) are not immediately binding for the board. Indeed, according to Section 4.7 of the new bylaws, "if the Board refuses or fails to comply with a duly authorized and valid EC Decision," the EC can "initiate a mediation process" with the board. In sum, the EC's capacity to constrain the board's behavior seems far weaker than that previously exercised by the NTIA.

Furthermore, the new set of reviewing processes, namely the IFRs and the IANA Naming Function Separation Process, are clearly insufficient in terms of accountability. In both cases, as we have seen, recommendations made by the IFR teams and the SCWG do not bind the board's decisions. Also, both these teams are not external to ICANN and are dominated by ICANN's contractual partners in terms of composition. In sum, the transition did not fix the lack of mechanisms "sufficiently independent of the ICANN Board of Directors and binding on it to qualify as truly accountable" (Weber and Gunnarson 2012: 21), as they had already emerged during previous accountability and transparency reviews.

---

[1] See p. 14 of the first draft proposal available at https://community.icann.org/pages/viewpage.action?pageId=52897394. Accessed 12 June 2020.

Even the PTI and the CSC are not external entities. In fact, while the PTI is a separate legal entity, ICANN is the sole member that nominates its board. The CSC is made of representatives of ICANN's direct customers, particularly gTLD's and ccTLD's registries. Hence, the CSC review of the PTI's performance may, at the best, improve only ICANN's internal accountability, and does not change its accountability toward external stakeholders and interests.

Concluding this section, the institutional effectiveness of the IANA transition cannot be evaluated as satisfying from a normative point of view in terms of inclusiveness, balanced representation, and accountability. As a consequence, the ICANN board remains the expression of interwoven business and technical interests and is unlikely to be truly constrained by an independent entity.

## 7.3 Outcome Effectiveness: A Still Contested DNS Regime

In this last section of this chapter, we address some controversies relating to the DNS regime as they have emerged in the post-transition period. This analysis aims to understand whether tensions traditionally related to the DNS and ICANN governance structure have been overcome or if they have re-emerged after the 2016 transition. The analysis focuses on two crucial cases that produced a public debate at the international level: the attempts by the Russian government to isolate its network from the global Internet and the agreement between ISOC and a private equity firm for the sale of the .org TLD.

### 7.3.1    National DNSs: The 2019 Russian Sovereign Internet Law

The so-called Russian sovereign internet law is, in fact, a series of three amendments to the Federal Laws n. 126 of July 7, 2003 "On Communication," and n. 149 of July 27, 2006 "On Information, Information Technologies, and Information Protection." These amendments were introduced by the Federal Law n. 90 of May 1, 2019, and came into force on November 1, 2019.[2] This law was framed by an official press note of the Duma as a counteraction against the 2018 US

---

[2] http://publication.pravo.gov.ru/Document/View/0001201905010025. Accessed 12 June 2020.

Cybersecurity National Strategy, which included the Russian Federation among the players accused of using "cyber tools" to undermine the US economy, democracy, and intellectual property, and which promised tough punishment, including the use of the force, against cyberattacks.[3] The amendments have three main goals. First, to oblige Russian ISPs to install technical equipment for counteracting threats. This equipment is provided by the Roskomnadzor, whose official name is Federal Service for Supervision of Communications, Information Technology, and Mass Media. The second provision of the 2019 amendments is that, in case of threats, the Roskomnadzor will activate a centralized control system over foreign interconnections and cross-border data routing. The third amendment concerns the implementation of a Russian national DNS, which will be mandatory for ISPs starting from January 1, 2021. All these amendments, and particularly the third, indicate that one of the most relevant controversies around ICANN and the DNS, the role and powers of the US government, has not diminished following the IANA transition. Notwithstanding the relinquishment of the US Department of Commerce's oversight over ICANN and the end of the NTIA's power to authorize changes in the DNS root zone,[4] there is no sign that the transition has improved trust and rule compliance among those actors, such as the Russian government, that have traditionally expressed dissatisfaction with the DNS regime that emerged at the end the 1990s. The Russian law may have been determined by other factors, such as the aim of the Russian government to implement more effective control and surveillance systems to target domestic dissent and opposition. On the other hand, it is understandable that, from the Russian perspective, as well as from the points of view of other US rivals such as China, the DNS remains too much tied to the USA due to the locations of ICANN and Verisign, as well as of the root servers operators, which are all based in the USA except for three that are located in allied countries (Japan, Netherland, and Sweden). However, if one of the main outcomes expected by the internationalization of the DNS was to avoid the risk of fragmentation and to keep the Internet as a unified space based on a common set of unique identifiers, it is clear that the transition has not solved this long-standing problem, nor has it lessened the geopolitical tensions relating to key Internet resources. The aim

---

[3] https://doi.org/http://duma.gov.ru/news/44551/. Accessed 12 June 2020.

[4] Amendment 33 of October 20, 2016, to the Verisign Cooperative Agreement.

of the Russian government to be independent of ICANN in addressing systems and domain names shows a clear rejection of the IANA transition's outcomes as a legitimate and ultimate solution from the side of an important actor in global Internet governance. This actor has expressed its aspirations for a national Internet since at least 2014 (cfr. Mueller 2017: 51), and those aims were not prevented by the transition.

### 7.3.2 The Territorial Jurisdiction of ICANN: The Case of the .org TLD

After the transition, the narratives of the completed privatization of the DNS and of ICANN being finally released from its embarrassing relationship with American authorities suffered a serious blow with the recent case of the .org TLD. The .org is one of the first six top-level domains created in the DNS space, together with .mil, .com, .edu, .govm and the temporary .arpa (Postel and Reynolds 1984). As seen in Chapter 3, since the early 1990s these domains were managed by Network Solutions, Inc. (NSI), which assigned the respective second-level domains, initially for free then charging a fee from 1995 onward. In 2002, at the end of a bidding process, ISOC established the Public Interest Registry (PIR) as a Pennsylvania not-for-profit corporation to take charge of managing the .org in 2003.

On November 13, 2019, PIR announced that ISOC had reached an agreement with the private equity firm Ethos Capital for the sale of the .org for 1.135 billion dollars. The day after, PIR formally submitted a "Notice of Indirect Change of Control and Entity Conversion" (a Change of Control Request) to ICANN in advance of closing the proposed transaction. From that moment on, a close correspondence between ICANN and ISOC began, consisting of a dense exchange of requests for information and documented details on the transaction. The proposed deal also raised a number of concerns. On November 22, 2019, a coalition of 27 civil society associations, later joined by hundreds organizations, wrote an open letter to the ICANN president CEO, urging him "to stop the sale of the Public Interest Registry (PIR) to Ethos Capital."[5] The main thesis advanced by the letter was that the .org TLD has been historically used by non-governmental organizations that have

---

[5] https://www.eff.org/files/2019/11/22/coalition_letter_on_pir_sale.pdf. Accessed 13 June 2020.

benefited from the not-for-profit nature of the PIR to promote the public interest. The asserted special nature of the .org, with its mission so deeply rooted in the concept of the public interest, was anchored to the legal instruments used throughout the 2002 bidding process to transfer the domain to ISOC's PIR. Moreover, the relationship between .org and the public interest was grounded in a political narrative portraying the .org community as morally committed to preserving the non-commercial portion of the DNS space, as testified by the term chosen for the registry since its establishment—the Public Interest Registry. In December 2019, there were other stances against the proposed transaction, including one authored by the UN special rapporteur on the promotion of the right to freedom of opinion and expression and the UN special rapporteur on the rights to freedom of peaceful assembly and association. The two UN rapporteurs added a human rights narrative to the public interest discourse and highlighted that:

> The proposed deal raises serious questions about the ability of civil society organizations and other public-minded individuals and entities to continue to enjoy the space for the exercise of the rights to freedom of expression and association offered by the .ORG domain managed by the PIR.[6]

In January 2020, four US senators and two representatives built on the narrative of public interest by appealing for ICANN to reject the proposed "Ethos Capital takeover of the .ORG domain":

> Public interest should be at the forefront of any ICANN decision, but it should be especially so in determining who should be approved to operate the .ORG registry. The Ethos Capital takeover of the .ORG domain fails the public interest test in numerous ways: it threatens the quality and reliability of .ORG websites, and could severely limit access to these domains via price increases and arbitrary censorship.[7]

The ICANN At-Large Advisory Committee and the ISOC Chapter Advisory Council Steering Committee joined the group of opponents. At the end of February 2020, the PIR and Ethos Capital sought to regain trust

---

[6] https://doi.org/https://www.icann.org/en/system/files/correspondence/kaye-voule-to-marby-20dec19-en.pdf. Accessed 13 June 2020.

[7] https://www.warren.senate.gov/imo/media/doc/2020.01.16%20Letter%20to%20I CANN%20about%20sale%20of%20.ORG%20registry.pdf. Accessed 13 June 2020.

and legitimacy with the .org community by voluntarily adopting a legally-binding amendment to the .org registry agreement, including a public interest commitment (PIC). A public engagement period was opened for ten days starting from March 3, 2020, during which the PIR received comments on the PIC. On April 7, 2020, the PIR submitted an updated PIC for consideration, with changes addressing some of the questions posed by the .org community, such as the establishment of a "Stewardship Council" to oversee the PIR.

On April 15, 2020, the ICANN board chair and the ICANN president and CEO received a letter from the Office of the Attorney General of the State of California (CA-AGO) urging ICANN to "to reject the transfer of control over the .ORG Registry." The CA-AGO had already claimed authority over the decision on January 23, 2020, stating that it would have reviewed the proposal of transaction based on its statutory powers as the supervisor of "the operations of all not-for-profit entities in California, including ICANN." This affirmation of authority was reiterated in the April 15 notice:

> My office is committed to protecting California's and the public's interest in a properly functioning and accessible .ORG domain system. ICANN has long recognized the unique nature of the .ORG registry as the Internet's home for noncommercial entities and interests. ISOC and PIR are charitable organizations that are accountable to their community stakeholders and to the public at large. In contrast, a private equity firm is accountable only to its investors.[8]

A couple of weeks later, the ICANN board made two resolutions. The first rejected the PIR's change of control request (Resolution 2020.04.30.01) and the second excluded any prejudice to the PIR to resubmitting a new proposal in the case it "successfully achieves an entity conversion approval in Pennsylvania through the Pennsylvania Court" (Resolution 2020.04.30.01).

Even if the ICANN board wanted to clarify that the CA-AGO's letter "does not alone determine or require" withholding consent from the requested change of control, several points support the hypothesis that it played a crucial role in the decision. First of all, the risks that the transaction would have affected the freedom of the .org community were already present, and accepted by ICANN, within the changes to

---

[8] https://www.icann.org/en/system/files/correspondence/becerra-to-botterman-marby-15apr20-en.pdf. Accessed 13 June 2020.

2019 .org registry agreement that allowed the PIR to increase registration fees for .org second-level domains without the approval of ICANN, to implement mechanisms for intellectual property protection, and to suspend domains used for illegal activities. Second, the rationale for the two resolutions, published together with the board's decision, included recognition of the reviews of the proposed transaction made by the California and Pennsylvania authorities. Indeed, these considerations by US public authorities were the final words in the document, being published just before the conclusions. Also, the second resolution, which refers to the possibility that the PIR could re-submit the request provided that it is enabled by the Pennsylvania court to change its nature from a not-for-profit to a for-profit entity, supports the authorization role of the US judiciary recognized by ICANN.

Accepting the authority claim advanced by the CA-AGO, ICANN passed from the NTIA's oversight to the supervision of the California public authorities, showing that the relationship between the DNS regime and the USA is still effective. This situation, on the one hand, may foster recriminations by those governmental actors that question the effectiveness of the transition in unbinding ICANN from its ties to US authorities. On the other hand, this situation offers new viable instruments to civil society to make their voices heard, notwithstanding that the IANA transition process has left them as weak as they were before. In both cases, the .org case testifies that the outcome effectiveness of the IANA transition cannot be considered high, regardless of the possible positions concerning the issue at stake.

Concluding this section, the analysis of the output legitimacy of the IANA transition, considered both as institutional effectiveness and outcome effectiveness, presents no evidence of improvement of the IANA stewardship in terms of inclusiveness, balanced representation, and accountability. The new institutional design of ICANN has not improved the inclusiveness of the organization. It is in line with the gradual strengthening of the GAC's power but does not make representation more balanced. Also, the transition failed in replacing the NTIA's oversight with any effective mechanism to hold ICANN accountable to an independent entity. Instead, the analysis clearly shows that the IANA transition has transmitted all the structural limitations and legitimacy deficits from the previous DNS regime to the new arrangements. The IANA transition resulted in a missed opportunity to overcome those legitimacy deficits already discussed and assessed during almost two decades of debate over DNS governance. Moreover, the transition failed in the attempt to define the "global Internet multistakeholder community" in a more universalistic and democratic manner, with meaningful inclusion of civil society, end-users, and developing countries, rather than just the "Internet people" already involved in the ICANN constituencies. This, in its turn, limited the possibility that the transition could solve, or at least reduce, DNS-related conflicts and tensions among global stakeholders.

## References

Bäckstrand, K. (2006). Democratizing Global Environmental Governance? Stakeholder Democracy after the World Summit on Sustainable Development. *European Journal of International Relations, 12*(4), 467–498.

Bäckstrand, K., Khan, J., Kronsell, A., & Lövbrand, E. (2010). *Environmental Politics and Deliberative Democracy: Examining the Promise of New Modes of Governance.* Cheltenham, UK: Edward Elgar.

Hofmann, J. (2016). Multi-Stakeholderism in Internet Governance: Putting a Fiction into Practice. *Journal of Cyber Policy, 1*(1), 29–49.

ICANN. (2016). *Bylaws for Internet Corporation for Assigned Names and Numbers.* https://www.icann.org/resources/pages/bylaws-2016-09-30-en. Accessed17 March 2020.

Mueller, M. (2017). *Will the Internet Fragment.* Cambridge: Polity Press.

NTIA. (2014). *Intent to Transition Key Internet Domain Name Functions.* https://www.ntia.doc.gov/press-release/2014/ntia-announces-intent-transition-key-internet-domain-name-functions. Accessed 12 June 2020.

Postel, J., & Reynolds, J. (1984). *Domain Requirements, RFC920.* https://tools.ietf.org/html/rfc920. Accessed 12 June 2020.

Weber, R. H., & Gunnarson, S. R. (2012). A Constitutional Solution for Internet Governance. *Columbia Science & Technology Law Review, 14*, 1–71.

# CHAPTER 8

# Conclusion: The Misleading Rhetoric of Multistakeholderism

**Abstract** This chapter summarizes the findings from the previous analyses and discusses the overall legitimacy of the IANA transition process according to the selected normative criteria. The results suggest that despite the IANA transition's success in removing US government oversight, it neither consisted of nor produced an improved model of multistakeholder governance. Based on the IANA transition case, the chapter concludes that multistakeholderism risks resulting in misleading rhetoric that legitimizes power asymmetries, rather than being a performative concept leading toward the democratization of transnational policy-making. Finally, the chapter calls for a reform of multistakeholder governance toward a model of digital constitutionalism.

**Keywords** Multistakeholderism · Internet governance · Power · Digital constitutionalism

This study investigated the legitimacy of the IANA transition process through a set of normative criteria commonly employed for the assessment of input, throughput, and output dimensions of policy processes and governance arrangements. The general aim of the analysis was to

© The Author(s) 2021
N. Palladino and M. Santaniello, *Legitimacy, Power, and Inequalities in the Multistakeholder Internet Governance*,
Information Technology and Global Governance,
https://doi.org/10.1007/978-3-030-56131-4_8

understand whether the IANA transition was able to overcome the well-known limits and bias of the DNS regime and whether it could be considered a significant improvement of the multistakeholder model in the Internet governance field.

Chapter 5 analyzed the input legitimacy dimension, in turn broken down into three criteria: inclusiveness, balanced representation, and representativeness. The findings point out that, even though the IANA transition process included all the relevant stakeholder categories, it was far from meeting a satisfactory standard of input legitimacy. Indeed, almost all the voting members and participants were affiliated with one or more of the structures within ICANN, the Internet Engineering Task Force (IETF), and the RIRs, failing to meet the widespread request for a process going "beyond ICANN" that was advanced during the preparatory phase. Further, the analysis sheds light on how the IANA transition process has reproduced the well-known power imbalances and misrepresentations of ICANN and the DNS regime. The data show an overwhelming presence of Western registries and private sector actors in the drafting bodies, while civil society and the Global South went largely underrepresented. Moreover, affiliation analysis highlights how revolving doors and multiple affiliations blurred the boundaries among stakeholder categories, raising the question of which interests and concerns the voting members of the drafting bodies actually represented.

The analysis of throughput legitimacy also returns a negative assessment as the process seems to have been flawed in many regards. The decision to entrust ICANN with the task of convening the IANA transition process put ICANN itself in a clear conflict of interest. ICANN's moves were contested since the beginning, and the suspicion that it wanted to take advantage of its role to steer the process toward its favored outcome soon emerged, even from inside the ICANN community. Despite the board reacting to these widespread allegations with a conciliatory attitude, in the end, it established an IANA Stewardship Coordination Group primarily based on its existing stakeholder structures and its contractual partners. In turn, the ICG entrusted ICANN, the IETF, and the RIRs to draft partial transition proposals on behalf of the name, protocol, and number communities, respectively. Actors not belonging to these organizations were allowed to take part in the discussion, but they did so with minor participatory rights and according to engagement rules established without their contribution.

Another point of concern was accountability. As we have seen, the chartering organizations of the CWG-Stewardship, the CCWG-Accountability, the CRISP Team, and the IETF retained approval power over the outcome of the respective working groups. In so doing, they established an accountability relationship among the people that materially drafted the transition proposal, the agent, and a broader community, the principal. However, this community was restricted, even in this case, to ICANN constituencies, the IETF, and the RIRs.

Nevertheless, this accountability mechanism is the first step in a longer accountability chain that involved the ICANN board, the NTIA, and the US Congress. Even though none of these subjects rejected or requested changes to the transition proposal, the analysis of the discussion within the CWG-Stewardship testifies that these accountability mechanisms influenced the behavior of participants, who took into account the preferences and expectations of the NTIA and the US Congress to prevent potential reasons for rejection.

The analysis of the discursive quality also reveals that, on the one hand, the deliberation around the IANA transition was respectful, rational, and animated by a constructive attitude. However, on the other hand, the discussion was inscribed within a neoliberal framework. Indeed, the overall institutional design of the process and the composition of the working groups reproduced the assumptions, values, and ways of thinking embedded in the organizational culture of the NTIA, ICANN, the IETF, and the RIRs into the deliberative process. The analysis shows that a neoliberal discourse was dominant, shaped conversations and policy options, and inhibited other discourses on the IANA and the DNS that are present elsewhere in the public debate over Internet governance.

Output legitimacy relies on the capacity of a decision-making process to reach the goals for which it was established. In the case of the IANA transition, this means assessing whether the process moved IANA stewardship from the NTIA to the global multistakeholder community and, more generally, the extent to which the traditional deficits of legitimacy affecting ICANN and IANA functions were effectively reduced. Without a doubt, the transition has brought into effect novel governance arrangements, removing the historical oversight of the USA. It is more questionable, however, if stewardship has moved to the global multistakeholder community and if effective oversight mechanisms are still in place. The final proposal provided for a new IANA functions operator, the Post-Transition IANA (PTI), established as an affiliate (subsidiary) of ICANN, which is the sole member of PTI and appoints its directors. The IETF and the NRO decided to maintain ICANN as the IANA functions operator for the protocol and number functions, allowing ICANN to subcontract their parts of the IANA functions to the PTI.

In the end, rather than transferring the stewardship of IANA functions to a new multistakeholder body that controls the IANA operator (ICANN), the transition process allowed the ICANN multistakeholder community to perform the oversight role that once belonged to the NTIA over a new IANA operator, the PTI, separated but incorporated within ICANN itself. ICANN, IETF, and NRO settled up standards for the IANA services. Nevertheless, only ICANN could exercise oversight over the IANA function operator, being the sole member of the PTI. In truth, these arrangements are likely to blur the boundaries between the operator and its supervisor. By removing NTIA oversight, the transition also removed the accountability mechanism based on the possibility that the NTIA could re-open the IANA Functions Contract for competitive bids, which has shown to have forced the ICANN board to reconsider some of its decisions in the past. In this model, the internal model, accountability rests solely on ICANN's governance mechanisms.

As seen in Chapter 7, the powers of the Empowered Community are hard to exercise effectively due to an overly complicated escalation and threshold system. In the end, the ICANN board is unlikely to be truly constrained by internal mechanisms and remains mostly the expression of interwoven business and technical interests. Instead, as the.org case testifies, US public authorities still have significant influence over the board's decisions.

The assessment of the extent to which the IANA transition complied with normative standards of legitimacy depends on what one means by "Internet global multistakeholder community." For those who believe that the IANA is a business concerning exclusively or primarily ICANN, the IETF, the NRO, and their respective communities, the IANA transition process could be considered inclusive and fair enough, and its outcome effectively transferring the stewardship over IANA functions to the global stakeholder's community of reference. For those who believe that the IANA stakeholders extend far beyond the organizations mentioned above, the assessment can only have a negative result.

In the end, we consider that the boundaries of the Internet global multistakeholder community cannot be limited to the ICANN-IETF-NRO system, and therefore, we conclude a more pronounced negative assessment. The rationale for this consideration relies on a set of interwoven reasons.

First, the IANA functions are at the heart of the DNS and the Internet as we know it. Thus, their governance and performance affect a vast range of actors that should be considered legitimate stakeholders. There was no reason to limit participation to just those organizations already having an operational relationship with the IANA.

Second, as testified by the final statement of the NetMundial meeting, by all accounts one of the most inclusive and fair Internet governance multistakeholder initiatives, the widespread opinion about the transition was that "the discussion about mechanisms for guaranteeing the transparency and accountability of those functions after the US Government role ends, has to take place through an open process with the participation of all stakeholders extending beyond the ICANN community,"[1] a position that was recalled in numerous contributions submitted during the preparatory phase of the IANA transition process.

Third, the mechanism through which the "internal model" was imposed is itself a striking example of the rhetorical use of the multistakeholder discourse. In particular, our analyses show that, through a neoliberal discourse, the key organizations already involved in the DNS regime were able to use the ambiguity of the concept of a "global multistakeholder community" as a strategic power resource. The origin of this ambiguity is the NTIA announcement itself, which seemed to define the global multistakeholder community in a twofold manner. The first part declared the intention to transition key Internet domain name functions to the "global multistakeholder community" and asked ICANN to convene "global stakeholders" to develop a proposal, which is a very general and potentially broad conception of the global multistakeholder community. Later in the announcement, the NTIA specified that "in the development of the proposal, ICANN will work collaboratively with the directly affected parties, including the IETF, the Internet Architecture Board (IAB), the Internet Society (ISOC), the Regional Internet Registries (RIRs), top-level domain name operators, VeriSign, and other interested global stakeholders."[2]

---

[1] http://netmundial.br/wp-content/uploads/2014/04/NETmundial-Multistakeholder-Document.pdf. Accessed 12 June 2020.

[2] https://www.ntia.doc.gov/press-release/2014/ntia-announces-intent-transition-key-internet-domain-name-functions. Accessed 12 June 2020.

In this last version, NTIA outlined a narrower—even if not exclusive—set of "special" actors, which would later be identified as the operational communities entrusted to draft the transition proposal. The ICG adopted the same scheme when it decided to entrust the "operational communities" to draft the proposal. This decision drew on a document submitted by the IAB that explicitly indicated ICANN, the IETF, and the NRO as the operational communities. Later, when the ICG launched a request inviting operational communities to submit their proposals, it indicated that the operational communities were "those with direct operational or service relationships with the IANA functions operator, in connection with names, numbers, or protocol parameters."[3] These decisions were promptly followed by the establishment of ad hoc working groups by the ICANN constituencies, the IETF, and the RIRs.

In sum, even if the call was formally open, its addressees were already identified as specific organizations. It is worth noting that these organizations did not involve external actors in the set-up phase. Rather, they only allowed other interested parties to take part in the discussion according to their rules and with minor participatory rights. In all of these cases, the reference to a broad and undefined global multistakeholder community served to strengthen the legitimacy of the process, rhetorically emphasizing openness and inclusiveness, while leading roles and decisional powers were, de facto, assigned to particular organizations, steering the development of the process toward what seems to be a predefined outcome.

The outcome of the process satisfied the preferences of the most powerful actors in the Internet governance field and solidified the dominant position of the organizations already carrying out DNS management (ICANN, the IETF, and the RIRs). After the transition, their control over the DNS has become more stable. In particular, ICANN was able to become the permanent holder of the IANA stewardship, combining policy-making and technical-operational functions, free of any external control. Since internal accountability mechanisms do not seem to be as effective as the previous NTIA oversight, the ICANN board and staff gained more space for autonomous action.

In the end, the novel governance arrangements strengthened the position of the registries and the technical community (which are often

---

[3] https://www.icann.org/en/system/files/files/rfp-iana-stewardship-08sep14-en.pdf. Accessed 12 June 2020.

interwoven, as we have seen in Chapter 4). Governments also acquired more power than in the past, while civil society confirmed its structural weakness within the ICANN governance structure. Nevertheless, this was an optimal result also for the US government, which was able to avoid the internationalization of IANA functions management, and relocation of ICANN itself, after the distrust produced by the Snowden disclosures. Since ICANN, the PTI, and most of the root server organizations remain on US territory, and therefore under US jurisdiction, the US government may still rely on a set of "last instance" sources of power. In the end, by taking the initiative, the US administration was able to prevent its geopolitical competitors from seizing the leadership of the DNS regime's reform, and maintained an exclusive, even if residual, point of advantage.

These latter considerations allow us to draw some conclusions on the overall meaning of the IANA transition process for the development of multistakeholderism in the Internet governance field. The IANA transition has been commonly described as a milestone in Internet governance. There is no doubt that it produced relevant changes to DNS governance, which in the last instance could be summarized as the sunset of an intergovernmental perspective in the short term, and in the finalization of the long-standing project to privatize the DNS pursued by the US government.

Nevertheless, our analysis suggests that the transition did not result in, nor did it lead to, a higher form of multistakeholderism filling the gap between reality and the ideal-type of what multistakeholderism ought to be, according to normative standards of legitimacy. Nor was it able to fix the well-known limitations in inclusiveness, fairness of the decision-making process, and accountability of the entire DNS regime.

Of course, the management of the DNS is now entirely carried out by a multistakeholder community. However, the type of multistakeholderism we have seen at work during the transition process, and even more the one resulting from this process, seem far from working as a "performative" narrative (Hofmann 2016) capable of guiding people toward the democratization of a transnational sphere of governance and the achievement of improved consensual outcomes. Instead, the transition seems to have solidified previous dominant positions and ratified the ownership of an essential public function by a private corporation, led by interwoven economic and technical interests.

In many regards, the IANA transition process and its outcome are a continuation of previous ICANN multistakeholder practices, which have been reinforcing "existing power dynamics" rather than "substantially redistributing power" (Carr 2015: 650). On the one hand, the transition seems to have resulted in a bargain between tech private firms and national governments. Governments, or at least the dominant coalition within the GAC, have given their consent to the definitive affirmation of the ICANN/IETF/NRO system in return for increased power in the novel arrangements, completing the ICANN's long-standing strategy of state engagement in its governance structure to gain international recognition and to prevent a shift of power toward the ITU or other intergovernmental bodies.

On the other hand, the transition process showed the irrelevance of civil society, little and badly represented in the stakeholder structure before and after the transition. In particular, despite some eminent representatives of civil society playing a relevant role during the deliberations, civil society as a stakeholder group was unable to speak with a clear and loud voice when chartering organizations approved the transition plan. According to our analysis, this was an effect of the marginality of civil society in the GNSO, as well as the affiliation with the private sector and technical organizations of most of the members appointed by the ALAC.

In the case of the IANA transition, multistakeholderism seems to have resulted in misleading rhetoric legitimizing power asymmetries embedded within the institutional design of DNS management, rather than in a new governance model capable of ensuring the meaningful participation of all the interested parties. This point calls for reflection on the relationship between normative and sociological legitimacy in the IANA transition case. Clear data on the perceived legitimacy of the IANA transition and the novel arrangements it produced are missing. ICANN collected numerous supporting statements from the US Internet community submitted to the Congress during its scrutiny of the transition proposal. The vast majority of these actors affirmed that the proposal preserved the stability, security, integrity, openness, and resilience of the Internet. Also, almost all of the statements authored by civil society organizations stressed that the transition plan would be the best way to prevent an intergovernmental takeover of the DNS management. Only a few business community contributions argued for the enhancement of the accountability of the DNS management. No one said that the novel arrangements would have improved the multistakeholder model toward a more inclusive, equal, and participatory governance setting. No one, again, advanced hope about the democratization of this policy domain.

A recent inquiry (Jongen and Scholte 2019) indicated that ICANN enjoys a moderate degree of confidence comparable to other, and often contested, international organizations such as the World Trade Organization, the World Bank, the International Monetary Fund, and the United Nations. However, considering that multistakeholderism is supposed to provide transnational governance with greater legitimacy than traditional intergovernmental modes, it can be said that in the case of ICANN, multistakeholderism did not fulfill its goal.

Even if future research detects increased confidence in ICANN, this would not be inconsistent with the conception of multistakeholderism as misleading rhetoric. Sociological legitimacy is about perceptions and does not necessarily rely on democratic values. Multistakeholderism may then result in a series of PR or stakeholder engagement practices (a sort of "participatory-washing") aimed at legitimizing existing power positions and dynamics without actually giving up power and without leading to a more fair, equal, and representative decision-making process.

Regardless of the degree of sociological legitimacy, a relevant question concerns how long a multistakeholder governance setting can survive and be effective without complying with normative standards of democratic legitimacy. In the last twenty years, ICANN, the IETF, and the RIRs have obtained remarkable results in the management of the DNS's technical functions, but numerous tensions, shortages, and deadlocks have occurred concerning policy-making and geopolitical issues, so much so that the fragmentation of the system has become a looming threat. As we have seen in Chapter 7, novel arrangements do not seem to have increased ICANN's capability to face the political issues at stake in global Internet governance.

Moreover, the analysis carried out in this book led to questioning the multistakeholder nature of the IANA transition. Avri Doria (2014) has already noted that the use of the term "multistakeholder" within the Internet governance ecosystem has often been an abuse masking the single stakeholder nature of many organizations and initiatives, which exclude other participants from real decisional power. The case of the IANA transition is quite different but leads to a similar conclusion. Findings in Chapter 5 reveal that IANA transition voting members experienced multiple and trans-sectoral affiliations, blurring the boundaries among stakeholder categories. Further, findings in Chapter 6 point out that a single discourse shaped the deliberation. These results contradict the assumptions at the basis of the multistakeholder model of governance, which is supposed to reach a higher and more complete understanding of a particular matter through deliberation among different categories of actors, with different backgrounds, views, and perspectives.

Instead, the set of IANA transition voting members in many regards resembled what has been defined as a "club governance" model, which refers to an "elite community where the members are motivated by peer recognition and a common goal in line with values, they consider honorable" (Tsingou 2015: 230). These observations are in line with previous studies that have described the community around Internet institutions as cross-sectoral and transnational power elite (Mueller 2010: 217–218) and multistakeholderism as a discursive tool putting them together (Chenou 2014).

The problem here is that governance settings, such as those described as club governance, base their legitimacy form professional expertise and reputation. They are well-suited to performing some form of "technocratic" governance, addressing an issue with a problem-solving approach based on an already given understanding of the nature of the problem and of the goals to be reached. Sharing a set of overlapping and compatible views is the cue that puts together these networks of experts. Nevertheless, they are ill-suited for tackling political problems, which, by definition, deal with pluralism.

It is worth recalling that the IANA was, first of all, a political matter. Indeed, the transition was settled as a consequence of a political fact—the widespread loss of trust in the USA as the caretaker of the Internet after the Snowden disclosures. Further, the IANA transition process aimed to achieve eminently political goals, such as establishing a novel governance setting and strengthening the DNS's accountability and legitimacy. Of course, IANA functions involve technical issues, nevertheless "they do not have only a technical dimension; they invoke some important and deeply-held values far removed from the 'merely' technical" (Post and Kehl 2015: 20). Around IANA functions there is a definitional struggle over the nature (public/private) of these global resources and about the proper governance setting.

In this context, a multistakeholder process was supposed to represent a diversity of views on the basic assumptions about the IANA functions, that, only after a highly political confrontation, can come to a common frame. Of course, during the transition, there were many discussions. However, they avoided the political aspects of the controversy around the IANA, even if they produced political effects. Chapter 6 shows how the shared beliefs on the technical and clerical conception of IANA functions played a crucial role in shaping the definition of oversight and stewardship.

As a consequence, the political problems connected to the IANA functions have been left unresolved, and, as shown in Chapter 7, it did not take a long time before they re-emerged. Moreover, the multistakeholder rhetoric has given to the new arrangements a veneer of democratic legitimacy, which could prevent contestation and the future development of a more equal and durable solution.

In this regard, the IANA transition does not appear as an isolated case. Initiatives supposed to be multistakeholder have often been criticized for not complying with their premises, resulting in "de-politicization mechanisms that limit political expression and struggle" (Moog et al. 2014: 6), which render "the implicit political choices invisible" (Cheyns and Riisgaard 2014: 7).

These findings lead to advance some suggestions for both future research and practice. Regarding the empirical side of research, this work highlights the need for a rigorous operational definition of a multistakeholder process to evaluate the cases that can meaningfully be included among multistakeholder initiatives. This study also emphasizes the relevance of taking into account the microdimension of policy-making, exploring individual data about the education and career of policy-makers, which can provide valuable insights about the nature of the process and may reveal hidden power structures. As seen, this data could shed light on the emergence of a transnational power elite, or community of policy experts around an issue, often transcending the borders of stakeholder groups and sharing common understandings based on similar educational or professional backgrounds, or from interactions in a transnational policy space (Tsingou 2015).

On the theoretical side, our findings call for the adoption of a more critical approach to the study of multistakeholderism. First, the more empirical research that confirms that multiple affiliations, revolving doors, and blurring boundaries among stakeholders are the rule rather than the exception among multistakeholder initiatives, the more scholars should question the validity of the concept of multistakeholderism itself. Its theoretical foundations, indeed, rely on the opportunity to bring to the table different categories of actors with distinct interests, knowledge, and expertise to reach a novel and more comprehensive view on the matter at hand.

When the differences among stakeholders are blurred, multistakeholder cooperation loses most of its meaning.

Further, the gap between the promises of multistakeholderism and its actual realization, highlighted by empirical research, should be put at the heart of the debate, and deeply investigated. Reflections should clarify if this observable discrepancy:

1. represents an unavoidable difference between an ideal-type and reality, and then determine what should be the acceptable margin of this gap;
2. is a problem of poor implementation, and then think about the institutionalization of possible remedies;
3. means that multistakeholderism is intrinsically flawed, structurally manipulable by the most powerful actors, and unable to offer adequate safeguards to the participatory rights of weaker stakeholders. In which case, it should be abandoned to direct efforts toward the development of alternative approaches.

As regards the Internet governance field, as Hofmann pointed out, multistakeholderism has become a mantra, a religion in which everyone who wants to have a say in the transnational Internet policy-making has to make a profession of faith. As she noted, "a measured 'desecration' of the multistakeholder approach in Internet governance which could facilitate a debate about achievements, failures and its reasons, would be a positive effect" (Hofmann 2016: 44). Internet governance studies could benefit from a more intense dialogue with the broader field of global governance (or with other fields of policy analysis, such as climate and environmental studies) where multistakeholder governance has already been widely investigated, with a more "secular" approach.

A frank debate on the potentials and pitfalls of multistakeholder governance should also impact the level of practice. Practitioners seem reluctant to discuss the shortcomings of multistakeholderism, probably because this discursive order provides them guidance and legitimation for their policy-making activities. Too often, this attitude has led to an uncritical celebration of multistakeholder initiatives, which, by masking poor results behind a thick blanket of rhetoric, has significantly slowed down the settlement of the many problems affecting the Internet governance field.

Meaningful advancements can be reached only by agreements on the definition of the problem. While multistakeholderism is used as a rhetoric to solidify and legitimize power positions within some policy-making arena, without any mechanisms giving up power to weaker stakeholders and without making concrete efforts to include different discourses, it will continue to produce ambiguous compromises without decisions, or make decisions affected by a poor degree of pluralism. In both cases, it will leave unresolved many of the problems and tensions it aims to address. Practitioners should be aware that the current fuzzy practices of multistakeholder governance in the Internet governance field are ill-suited to address the growing challenge posed by Internet development, and efforts should be made to re-think multistakeholder participation or elaborate innovative solutions.

Recalling an old but still relevant proposal advanced by Mueller, Mathiason, and Klein in 2007, the Internet governance field could benefit from the experience of the broader global governance field and look at the UN framework convention as a viable instrument. Framework conventions are "international agreement on foundational regime principles" (Mueller et al. 2007: 243), which "establish the principles and norms under which international action would proceed and set up a procedure for negotiating the more detailed arrangements" (ibid.: 251).

According to the authors, an Internet governance framework convention "should have agreed definitions for key facts or principles about the Internet and should clearly establish the norms that should be applied to Internet governance"; "indicate those areas in which further agreements need to be reached" (idem); as well as establish participatory rights for all involved parties to allow actors to take part in the decision-making process on an equal basis. As the experience of climate change policies testifies, framework conventions can be a viable instrument to provide multistakeholder agreements with the bindingness and safeguards of international law, taking the best of the two systems while minimizing risks.

In so far as such an instrument would be able to ensure the participatory rights of all the actors involved, sound anchorage to international human rights law, and enforceable rules, it could evolve multistakeholderism in a novel approach, or in other terms, it could constitutionalize multistakeholder governance. A digital constitutionalist model in this field could provide order through shared principles and norms, and societal control over the development of the Internet, constraining the power of both states and private operators. We are aware that such a perspective could sound utopian, but the disruptive transformations we are living through nowadays maybe require ambitious political projects.

## REFERENCES

Carr, M. (2015). Power Plays in Global Internet Governance. *Millennium: Journal of International Studies, 43*(2), 640–659.

Chenou, J. M. (2014). From Cyber-Libertarianism to Neoliberalism: Internet Exceptionalism, Multi-stakeholderism, and the Institutionalisation of Internet Governance in the 1990s. *Globalizations, 11*(2), 205–223. https://doi.org/10.1080/14747731.2014.887387.

Cheyns, E., & Riisgaard, L. (2014). Introduction to the Symposium: The Exercise of Power Through Multistakeholder Initiatives for Sustainable Agriculture and Its Inclusion and Exclusion Outcomes. *Agriculture and Human Values, 31*(3), 409–423.

Doria, A. (2014). Use [and Abuse] of Multistakeholderism in the Internet. In R. Radu, J. M. Chenou, & R. Weber (Eds.), *The Evolution of Global Internet Governance* (pp. 115–140). Berlin: Springer.

Hofmann, J. (2016). Multistakeholderism in Internet Governance: Putting a Fiction into Practice. *Journal of Cyber Policy, 1*(1), 29–49.

Jongen, H., & Scholte J. A. (2019). *Legitimacy in Multistakeholder Global Governance—Patterns at ICANN.* Paper presented at the GigaNet Annual Symposium.

Moog, S., Spicer, A., & Böhm, S. (2014). The Politics of Multi-stakeholder Initiatives: The Crisis of the Forest Stewardship Council. *Journal of Business Ethics.* https://doi.org/10.1007/s10551-013-2033-3.

Mueller, L. M. (2010). *Networks and the States.* Cambridge: The MIT Press.

Mueller, L. M., Mathiason, J., & Klein, H. (2007). The Internet and Global Governance: Principles and Norms for a New Regime. *Global Governance, 13,* 237–254.

Post, D. G., & Kehl, D. (2015). *Controlling Internet Infrastructure: The 'IANA Transition' and Why It Matters for the Future of the Internet, Part 1.* https://static.newamerica.org/attachments/2964-controlling-internet-infrastructure/IANA_Paper_No_1_Final.32d31198a3da4e0d859f989306f6d480.pdf. Accessed 15 September 2019.

Tsingou, E. (2015). Club Governance and the Making of Global Financial Rules. *Review of International Political Economy, 22*(2), 225–256.